John Read loves the dry inhabitants. Read his book and you'll understand why he does so, and you'll have fun along the way too.

Steve Morton, head of CSIRO Sustainable Ecosystems, *Nature Australia*, Autumn 2004

This book was a special treat — a story about a part of Australia few Australians understand ... an exciting read that made me feel like decamping to the outback.

James Woodford, *Sydney Morning Herald*

Read can obviously spin a good yarn around the camp fire. His book is a mixture of rollicking anecdote and ecological explanation, combining tales of his adventures ... with discussions of the delicate arid-zone environment of the desert around Lake Eyre.

Lorien Kaye, *The Age*

Red Sand, Green Heart is a book of anecdote, adventure and instruction. It is a piece of nature writing, accessible, rambling, lively and important ... The real desert is in there. So is our future.

Mark Tredinnick, *Canberra Times*

Red Sand, Green Heart is the rare insight of an ecologically educated writer with a flair for language and a passion for the environment.

R.M. Williams Outback Magazine

This is a very useful book as well as being an enjoyable read ... it should be read by everyone at all interested in conserving the Australian outback.

M. Kirton, *The Aurora*

RED SAND
green heart

ECOLOGICAL ADVENTURES IN THE OUTBACK

John L. Read

Lothian
BOOKS

Thomas C. Lothian Pty Ltd
132 Albert Road, South Melbourne, 3205
www.lothian.com.au

Copyright © John Read 2003

First published 2003
Reprinted 2005

All rights reserved. No part of this publication may be reproduced, stored in a retrieval system or transmitted in any form by any means without the prior permission of the copyright owner. Enquiries should be made to the publisher.

National Library of Australia
Cataloguing-in-Publication data:
Read, John L.
    Red sand, green heart.
    ISBN 0 7344 0478 6.
    1. Ecology - Environmental aspects - South Australia. I. Title.
333.72099423

Cover and text design by Studio Pazzo
Typeset by Studio Pazzo
Cover photograph by Peter Walton, Australian Scenics
Text photographs by J Read, J Fewster and T Lewis
Printed in Australia by Griffin Press

CONTENTS

Preface   xi

Part One   The desert lottery
1. How much did ya get?   3
2. Lessons from the trees   13
3. Every good drought is broken by rain   22
4. Don't bother with the forecast   28

Part Two   Just add water
5. Refugia   37
6. Stranded on a desert island   46
7. Climbing up a flat lake   52
8. Cameras can lie!   64
9. Giant white bees   77
10. Ducks in the desert   84
11. DUCKOFF   95

Part Three   Scales and tails
12. Life on the moon   107
13. Taipan country   121
14. How long have you got?   128
15. The climatic tightrope   141

Part Four If critters could talk
16 27km NE of Lyndhurst   157
17 Buckets of frogs   168
18 Outback canaries   177

Part Five Whose outback is it?
19 Who's shittin' who?   191
20 Soft lighting   207
21 Treading lightly   216
22 Cherish or perish   223
23 Gengelopes   237

Part Six Fighting off the ferals
24 More than meets the eye   257
25 Tracks in the sand   267
26 Cobwebs in crevices   277
27 Who's responsible?   288
28 Looking forward to the past 295

    Bibliography   311
    Acknowledgements   319

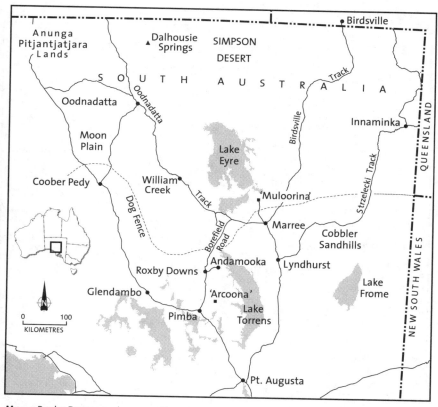

Map 1: Roxby Downs and surrounding areas

Legend

- Towns
- Homesteads

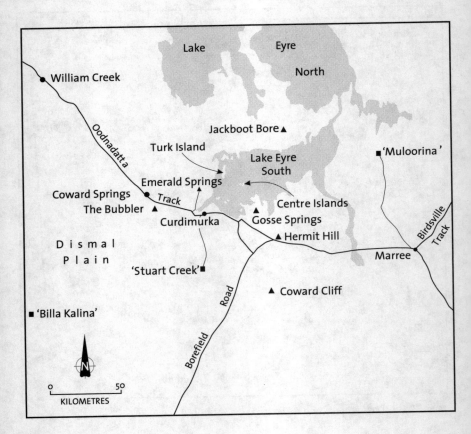

Map 2: Northern detail

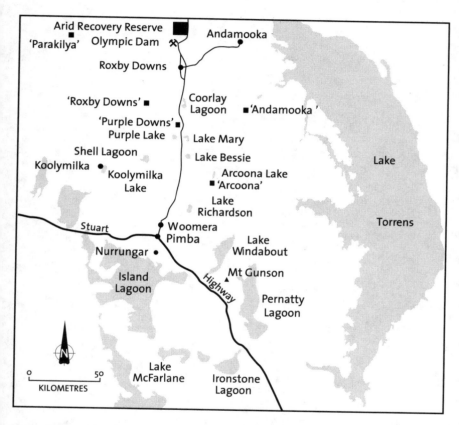

Map 3: Southern detail

Preface

In January 1989 the research centre on Kangaroo Island was buzzing with a group of enthusiastic Earthwatch volunteers. For the past week they had been assisting my ecological mentor, Dave Paton, to compare the effectiveness of native honeyeaters and introduced honeybees at pollinating native plants. Distractions were not what Dave needed while trying to paint coloured dots on live bees' bums without being stung or squashing his subjects. The phone rang but it was not for him. The call was a message for me to call Barry Middleton at Roxby Downs. Mum, who was relaying the message, sounded excited. I suspected he was offering me a job. I was torn.

For three years during the university holidays, I had monitored the rehabilitation of vegetation at the fledgling mine in the desert at Roxby. The mine was based on one of the world's largest copper orebodies and had just commenced production after a decade of environmental impact studies. Barry was the mine's new Environmental Superintendent. His offer of full-time employment threw me into a quandary. My primary interest on Kangaroo Island was not painted bees, but pygmy copperhead snakes. A mate and I had landed a World Wildlife Fund grant to study them over the following year. Venomous snakes were not a high conservation priority for many people and I knew that it was important to attempt to prevent their demise.

Prior to the call I had not given much thought to working full-time at Roxby. It was a great place to make some money over the Christmas break, to encounter arid zone wildlife and to enjoy the social highlights of a thriving country town in the silly season. Adam Smith, the incumbent ecologist who was leaving Roxby, told me that most of the interesting biological finds had been made. After a couple of years the job was losing its gloss. By accepting the Roxby offer would I be forfeiting a rewarding conservation research project for a boring job with money and security? Worse than that, would I be selling my soul to a mining company?

The Olympic Dam mine at Roxby is operated by WMC Resources, which used to be known as Western Mining Corporation, one of the largest Australian mining companies. Olympic Dam is principally a copper mine, but includes attractive resources of gold and silver as well. The ore also contains rich reserves of uranium, creating numerous environmental and health issues and raising the ire of the anti-nuclear lobby. By accepting a permanent position at Roxby, would I be condoning a mine or an industry that unduly compromised my principles?

I had been challenged by Julian Reid, a well-respected field biologist whom I worked with during his biological survey of the rich Coongie Lakes district. Julian was an opponent of the Olympic Dam mine, largely because uranium was being extracted. Like many people, Julian felt that the developers of the mine should be more open with the results of their extensive environmental monitoring. The place was shrouded in secrecy. Secrecy aroused suspicion. Because the environmental reports were not available to the public, the sceptics suspected that the miners were attempting to conceal the real impact of their activities.

I could sympathise with these views. When I first got my driver's licence I refused to fill my grandmother's 'borrowed' Hillman Minx at BP service stations. BP was a half owner of the Olympic Dam mine. This mine produced uranium. Uranium was radioactive and was used in bombs and hence was bad. In 1983 anti-Roxby protests were in full swing and the clashes between protestors and cops, both of whom outnumbered miners, had top billing on the Adelaide news reports. Boycott BP to close Roxby Downs. The anti-nuclear logic certainly influenced me as an impressionable teenager.

Was the huge Olympic Dam juggernaut hoodwinking the public? Was it achieving its pre-development claims of minimal environmental impact? My allegiances were biased toward the environment. If cover-ups and environmental degradation were rife, as some of my mates suspected, working at the mine would place me in a strategic position to expose them from the inside. If I wasn't morally outraged by what I found, maybe I could help the developing mine to maintain environmental values. Eventually I convinced myself that going to Roxby was a win/win scenario, and as a result the copperhead study was knocked off in a couple of hectic months, instead of twelve.

In March 1989 I packed my worldly possessions into my Belmont station wagon and headed north. It was an amazing feeling when I reached the real bush north of Port Augusta after driving through the agricultural land to the south. I felt like I was coming home. Colourful parrots flew across the road from the woodlands of myall and mulga trees, whereas further south I saw mainly flocks of feral starlings over cleared paddocks. Little earless dragons perched on rocks and watched me drive past them on the stony flats. Far from being the barren desert that is often perceived, this country seemed alive and exciting. Six hours and 560 kilometres after leaving Adelaide, I reached the newly built town of Roxby Downs.

Unlike fly-in fly-out company towns like Moomba, Roxby was open to the general public. The young families of mine workers lived alongside teachers, storekeepers and contractors who were lured by the boomtown opportunities. Roxby more closely resembled a modern subdivision on the outskirts of a major city than most country or mining towns. Although it lacked character, Roxby was an exciting young-persons' town, where grey hairs were as scarce as unemployed youth and vacancies on the brand-new basketball, squash and tennis courts.

On my third day at Roxby, after skimming through the environmental files, I sat down in Barry Middleton's office. Prior to this I had not been interviewed, no job specification had been prepared and no-one had indicated what my responsibilities were, except that I was 'The Ecologist'. Barry leaned forward and instructed me in a serious voice, 'Your job is to do the animals'. He then sat back and watched a rather puzzled expression distort my face. 'Do the animals?' I questioned, as I tried to get a grip on the scope of my new job. 'Yes that's right', Barry proclaimed. When I pushed for more details I was told that as an ecologist I should work it out for myself.

Environmental departments of major mining companies were traditionally staffed by geologists because they had an experience with, and an affinity for, the outdoors. However, in recent years the importance of environmental departments has increased exponentially and the qualifications of their personnel have changed. Most environmental staff are not the 'green' type who look at plants and animals but the 'browns' who monitor emissions and spills. These brown environmental scientists develop pollution control procedures, draft waste management proposals and produce mine closure plans. A dozen biologists, radiation experts, chemists, hydro-geologists, horticulturalists, field and support staff reported to Barry. I was fortunate to be at Olympic Dam because it had a high enough profile to justify a full time ecologist, just to 'do' the animals. Barry told me that Frank Badman, 'The Botanist', 'did' the plants.

※ ※ ※

Until a few decades ago ecology was considered to be a weak science largely restricted to describing how, why, where and when living things respond to other living things and their environments. Relatively recent advances, somewhat analogous to those championed by Newton and Einstein in physics, have made ecology a far more predictive science. A whole set of ecological $e=mc^2$-type rules have been established. Coupled with this increased sophistication in the understanding of ecology is a greater awareness of local and global environmental crises. Both have thrust this relatively young science into the forefront of management and political agendas.

In practical terms, an ecologist working for a mining company is a bit like a sun-tanned accountant. An Environmental Impact Assessment is an outdoor real-life budget of the predicted environmental consequences of a particular development. The ecologist's ledger is compiled using binoculars and traps. However, interpreting the results is not as simple as profit and loss.

In a wet year budgies will breed in dirty truck exhaust pipes and native mice populations will boom regardless of the environmental performance of a mining operation. Sometimes, no matter how environmentally responsible a mining company is, natural parasites will kill trees, feral animals will eat wildlife, and drought, floods and fires will maim individual animals or even destroy populations. It does not take great expertise to explain these more obvious responses of plants and animals to their environment. However, it is the subtle changes that demand more ecological nous.

It gets difficult if we don't know whether the frog with the extra leg has been affected by radiation or other pollutants, or whether it is just a rare yet natural freak. Are sulphur gases from the copper smelter responsible for the declining gecko numbers or is it increased predation? Research is required to answer the multitude of questions that arise when environmental changes are monitored. This long-term research into the lives and habits of largely unknown outback animals provided me with a rare opportunity for such scholarship.

Despite the opportunities offered by my new job, the pressure was on from the start. Even before I left for Roxby my mates were asking me questions:

'Does the Olympic Dam mine wreck the environment?'

'Are you going to help to shut it down?'

'Which endangered species live there, and why are they in trouble?'

Crikey! Not only did I not know the answers, I did not even know what questions to ask. First, I needed to understand the processes that shape the landforms, plant and animal communities before I had a hope of measuring impacts. Only after I had established how serious the impacts were could I decide whether the mine was 'good' or 'bad'. More importantly, only then would I be in the position to decide how best to minimise environmental impacts, or better still, how to help to improve the conditions for outback plants and animals.

But hang on, Barry had told me that I was 'The Ecologist' and I should know about these things. What's more, I was the only resident ecologist for hundreds of kilometres. On went my hat.

PART ONE
The desert lottery

1
How much did ya get?

White marbles shot from under the tyres as the road took me north, leaving behind a billowing cloud of bulldust. Ahead the white road cut through the shimmering desert where grey stunted bushes barely cast a shadow on the brick-coloured gibber stones. Sharp static cracks on the radio indicated lightning strikes somewhere between the broadcast at Port Pirie and my destination Lake Eyre. As I sped over a low dune the voices on my UHF radio merged with the clink of road gravel and the sounds of the *Country Hour* program on the car radio. A few years ago the conversation on my UHF would have seemed about as interesting as the cucumber prices that were being quoted on the radio. However the trivialities and repetition of the discussion on which I was eavesdropping camouflaged a far more important issue. I pulled up on the next dune and silenced the market reports as the white dust settled on the dash. A flock of zebra finches on the dead mulga tree was listening too.

'Oh yeah no worries then mate ... we got about eight points here last night but Malcolm said this morning that they got a fair bit more over Marree way, Muloorina got half an inch, Clayton thirty points, how much did ya get ... back.'

'Yeah righto Bob, got all that, no nothing much to report from here, we only got two points and Leo didn't get anything but Anna Creek got about sixty so the rain musta gone over the top of us here ... over the lake there somewhere ... we were watching the lightning out your way so thought you might have got a bit more around Wood Duck there ... bad luck about that ... your turn next eh?'

'Yeah no worries ... yeah those Marree and William Creek mobs seem to be getting it all lately ... we've only had just over two inches for the year, after only four last year, so things are getting pretty dry, the house dam is nearly finished ... ah no worries ... yeah that storm that you were watching was up past Coward there, so One Box might have got a bit of a drink ... no worries then ... so did you get a look at the weather map there this morning ... back.'

"Sa pity that storm missed you there ... there's that cyclone out off Hedland there which might shoot some rain over ... but I wouldn't hold my breath, I don't reckon it has been hot enough for a decent rain lately ... yeah we're pretty dry here too but not as bad as poor old Robby who has been carting water for months now ... anyway, got to meet the truck out at Boundary directly so I better get going now Bob ... so cheers.'

'Yeah ... no worries ... cheers.'

All around the north, on different UHF channels and phone lines, rainfall recordings were being compared and discussed. Telstra shares are worth buying simply because homesteads now have telephones and the hint of rain in the district evokes a rush of calls to find out who got what. Pastoralists, every bit as much as the zebra finches outside, are creatures of the bush. The lives of everyone and everything out here are shaped by the rain.

When it comes, the iron roofs amplify the pattering and drive everyone outside for a look. Kerosene grass seeds that have been impaled in the sand for months miraculously start screwing themselves downwards as the first drops hit. Deep breaths are taken, automatically sucking in the sweet smell of rejuvenation, and if it pelts down the excitement really sets in. Conservative adults act like red cordial-charged five-year-olds at a party.

In the cities rain is irrelevant. Stormwater drains gobble up the run-off along with leaves, iced-coffee cartons and cigarette butts. The memory of an inch of rain lasts as long as it takes for the Super Soppers to get the cricket or tennis back on schedule. In the hills rain is quickly whisked away down streams or soaked into the leaf-strewn forest floor. A day might be wet, a whole week might be wet, but when the sun rises on a cloudless dawn the rain is forgotten.

Rain is different out here and not just because it is typically scarcer than anywhere else in Australia. Most of the land is so flat that there are few rivers or even temporary watercourses. Rain usually stays where it falls. The few millimetres from last month is still evident by the tinge of green in the grass or the tyre treads of a previous vehicle fossilised into the track. The effect of a couple of inches can be seen for years and people reminisce about flooding rains for decades. Rain is everything out here. It is never convenient, always needed and perennially topical.

I continued down the road and through a low-lying basin sparsely vegetated by wiry, apparently lifeless canegrass. In 1989 this swamp held over a metre of water. The mirage that distorted the horizon through my windscreen was no illusion back then. Swans and ducks were nesting where my tyres now belched clouds of bulldust. The smell of luxuriant ephemeral growth was now gone. So too were the artillery of grasshoppers that battered the windscreen, the flocks of budgies that carved up the sky like schools of tropical fish and the crimson chats that expertly rode imaginary bow waves in front of the car.

Taking into account the erratic climate for which central Australia is renowned, 1989 was out of the groove. To be more precise: March 13 and 14 were climatic freaks. A monsoonal depression near Derby in Western Australia made its way along a surface trough into northern South Australia. With it came more rain than normally falls in a whole year. Much more. The figures being tallied from the Flinders Ranges to the outback had not been seen on rainfall charts for decades or, in some cases, for centuries. Motpena, in the Flinders Ranges, broke the rainfall record for the state with a whopping 273 millimetres in twenty-four hours. Further rains the next day increased the tally to 340 millimetres, over thirteen inches on the old scale.

For a change it was the outback that dominated the television stations that night. Livestock and wildlife were stranded on islands of high ground. Opal prices were predicted to soar as claims at Coober Pedy, Mintabie and Andamooka were flooded. Only the roof of the old Roxby Downs Homestead escaped the rising waters. Hundreds of kilometres of fences were flattened and dams were breached. Andamooka lost its entire water supply from the Chimney Hole Dam into Lake Torrens. Airstrips were under water and roads were cut. No-one went anywhere, but tourist operators were already spruiking about what a bonza season the blooming wildflowers and full lakes would bring.

Even the weather was changed by the big rain. The crisp dry heat of late summer was replaced by a thick suffocating humidity. One of David Attenborough's time-lapse cameras would have shown the same rain falling again and again, as the water was sucked out of the waterlogged landscape before bouncing off the clouds and back to ground again. The single blokes at Roxby remember these times well. Every Friday afternoon a testosterone-driven convoy of big-smoke pilgrims was thwarted, or should have been, when the only road connecting Roxby with Adelaide turned to mud.

※ ※ ※

All major rains, even those that are not as spectacular as the 1989 floods, stimulate an avalanche of wildlife responses. Even as the storm clouds build animals start responding. Occasionally a high-pitched staccato twang can be heard from the heavens as the wispy clouds stream in preceding a storm. Staring into the rushing grey clouds the silhouettes of whirling sickle-shaped birds with long swept-back wings eventually come into focus. As they race past I stare intently to establish whether their tails are long and tapered or relatively short and blunt. Are they fork-tailed swifts or white-throated needletails? Chances are that up here they will be swifts; needletails are more common down south.

Although swifts accompany many of the largest storm fronts that cross Australia from west to east, I have never seen them flying back again. Maybe these 'storm birds' have some sort of 'return-to-sender' system on the east coast, where they are collected at Byron Bay and sent back to Shark Bay in the extreme west of the continent?

If the rain that these storm riders precede is heavy enough to soak the ground and run off to low-lying areas, a mass evacuation commences. The first refugees are typically invertebrates that have dug their holes in flood-prone locations. The ones that draw the most attention are the bitey or stingy kinds. Arguably the most fearsome-looking critters are the black hairy spiders with enormous fangs which bear a striking resemblance to the related, yet much more venomous, funnel-web spiders of the east coast. Since some of these ground spiders, or myglamorphs, can live for twenty years or more we can assume that they have experienced floods like this before.

Scorpions may have to evacuate their spiralled holes and it is usually only after rain that they are seen during daylight. Unlike the potentially fatal stings from some scorpions from Mexico, India and Africa, the two common species in the South Australian deserts give a jab like a burning needle that usually leads to nothing more than a bout of dizziness. The local scorpions pass through at least six larval or instar phases, each lasting approximately a year before reaching adulthood. As a result of this protracted development they live longer than most local reptiles, mammals or birds, which is remarkable for a small invertebrate. Although I have a bit of a soft spot for long-lived spiders and scorpions that have evacuated their flooded holes, another one of their crawly mates really gives me the heebie-jeebies.

Scolopendra centipedes are found in many countries, which means that thousands of biologists have probably suffered the same experience as I have when plunging a hand into a pitfall trap designed to catch lizards and mice. Unlike ants, spiders, scorpions or wasps that bite or sting quickly, centipedes grab hold with all of their legs and refuse to be dislodged until they have chewed into your fingers with their menacing jaws. More than once I have done myself a nasty injury simply trying to extricate a centipede from my hand in a flurry of arms and legs. My typical post-rain nightmares consist of these demons zigzagging irregularly towards me so I never quite know which way to flee. Forget the movies where the luckless hero is banished to a dungeon of limp snakes—a room full of centipedes is a far more alarming vision of hell!

As the rain continues other subterranean critters also get stirred up. Crickets re-excavate their holes, pushing up turd-like cylinders of sand onto the surface. Winged termites and ants, all prepared for mating flights, emerge almost miraculously from their underground galleries. Some even build little launch towers of sand to assist them with their take-off. To take advantage of the water and bountiful supplies of displaced or sexually activated insects, burrowing frogs emerge from their underground slumber. While the frogs revel in the sodden conditions, listless reptiles wash or swim to the edges of flooded areas and wait for the sun to restore their energy.

The animals that appear during rain are just the precursors of a host of 'plagues' that follow a big rain. Prehistoric-looking shield shrimps and other small crustaceans hatch from drought-resistant eggs and race through their cameo lives before the ponds and claypans dry. Almost overnight, a myriad of different moths appear and start laying eggs on the fresh new shoots. First the little grey moths and then a night or so later the large, black and white hawk moths swarm over the vegetation. These hawk moths resemble little fighter planes complete with the red 'afterburner' patches on their lower wings. Weeks later, huge brown window moths have their couple of nights of glory, to the delight of black kites that swoop up their carcasses from roads the following mornings.

Hot on the tails of the grubs and caterpillars are the predatory carab beetles. One particular species known to entomologists as *Calosoma schayeri* is well-known to outback residents as the stink beetle. As well as being voracious predators of caterpillars and the producers of an indescribably pungent stench, these iridescent green–black beetles are renowned for their attraction to light. When the stink beetles are booming, campfires and house lights may attract hundreds, while industrial lights attract millions from the greening bush. Both the Roxby metallurgical plant and the Moomba gas fields typically get buried in deluges of crunchy green beetles that block drains and interrupt work for several days, a month or so after a big rain.

No-one holds the stink beetles in high regard. Even the beetle-eating lizards generally turn up their noses at the olfactorally obnoxious stinkers. However, memories of the stink beetles quickly fade when the plague locusts move through. When they are in low numbers locusts hop around unobtrusively nibbling on grass like other grasshoppers. But in favourable seasons when their populations increase these innocent grasshoppers form into devastating swarms. Physiological changes triggered by high densities cause the grasshoppers to take to the skies *en masse*.

All of a sudden huge swarms move through an area overwhelming their predators and consuming much of the flush of plant growth like a cold bushfire. On cool mornings locusts 'pour' from bushes onto the ground as masses of dopey hoppers leave their slumber spots for their breakfast in grassy patches. Once it warms up locusts take to the air in their trillions, all somehow coordinating to fly in the same direction. Tonnes of locusts can descend upon an area from over 500 kilometres away. If they find further green patches they may breed again, each female locust laying up to four pods of thirty to fifty eggs in a month before dying. Two to three weeks later the little nymphs hatch and they too can lay eggs after another four to six weeks.

Modern society finds it difficult to accept that nature throws up these plagues of biblical proportions, particularly when they can devastate livelihoods. Therefore hundreds of tonnes of an organophosphate insecticide known as fenitrothion are sprayed over huge tracks of pastoral and agricultural land to control outbreaks and protect crops. The Australian Plague Locust Commission (APLC) is charged with the unenviable task of preventing over $100,000,000 worth of agricultural damage that can result from a single outbreak. At the same time they must minimise the effect of spraying this broad-spectrum insecticide across thousands of square kilometres of native grass and shrubland.

Fortunately for cattle breeders and kangaroo harvesters, fenitrothion breaks down rapidly through exposure to water and sunlight, and two weeks is ample time to allow cattle and roos to purge the chemical from their system. The fate of native animals is not so clear. Fenitrothion is highly toxic to aquatic organisms and hence is not used near creeks or in areas where threatened fauna are known to exist. Frogs, crustaceans and insects that inhabit small swamps in sprayed regions will be knocked around seriously. What's more, despite their best intentions, neither the APLC or any other organisation knows the accurate distribution of threatened animals, particularly in the deserts where rare critters come and go with the seasons.

In an effort to reduce the environmental impacts of locust control the APLC have developed a fungus that is believed to control only locusts. Yet even the use of this fungus which should alleviate the concerns of most organic pastoralists is not without environmental consequences, as both the fungus and insecticide serve to dramatically lower the locusts' population. Consider the poor insectivorous marsupial dunnart that has just endured a dry spell and is licking its lips with the arrival of some favourite tucker. Or the sleek pratincole that has flown down from New Guinea to take advantage of the multitudes of insects in summer. If these predators along with others such as kites, kestrels, dragon lizards, goannas, and mammals are deprived of their locust feast, it is not only bad news for them but it is also bad news for the APLC. Depletion of the natural predators of this indigenous insect may exacerbate the next outbreak.

Although stink beetles are the most annoying and plague locusts are the most damaging of the insect plagues, the most consistent and long-lasting response following rain is from flies. It is little wonder that dorky-looking flynets become bestsellers in outback roadhouses soon after rain. Flies develop rapidly and can produce seven generations within the space of a year. If we assume that a fly lays 120 eggs per generation and half of these develop into females, a single female fly can give rise to over five million million flies within a year. That's right: five million million. And the scary thing is that this breeding frenzy may be taking place hundreds of kilometres away and that flies can be rapidly blown in by the wind. With such phenomenal reproductive potential we are heavily reliant upon harsh weather and the natural control of flies to keep life bearable. A significant reduction in predator numbers may exacerbate the fly scourge. At some stage we may have to accept nature's idiosyncrasies rather than try to control insect avalanches with chemicals.

Another insect blowing in on the wind after big rains is the mosquito. In recent years the number of cases of Ross River fever and other mosquito-borne diseases has increased. Occasionally however, mossies bring benefit to the outback by transmitting myxomatosis which can dampen the huge potential increase in rabbit populations following decent rains.

Later still in the series of post-rain plagues, outside lights and windows are soon bedecked with lacewings resembling a cross between a dragonfly and a moth. Lacewing larvae are fearsome spiky balls with a huge set of jaws. They bury themselves a few millimetres below the ground surface and flick out an inverted cone of sand or clay. Ants fall into these steep divots and some eventually slide to the bottom where they are seized by the lacewings' fearsome jaws. Their ferocious appearance and ant-based diet has earned these little critters the name of ant lion. Presumably, since the ants are not as dependent upon rainfall as are the caterpillars and beetle larvae, lacewings do not plague until after other insects have had their glory.

Insectivorous birds like chats and songlarks soon move in to take advantage of the insects. Seed-eaters like budgies and cockatiels fly in, and then other predators arrive to tackle the influx of birds. Within a few months the populations of native and introduced mice erupt, also buoyed by the abundance of seeds. The following year the snakes that have fed on the mice and bred successfully turn up more regularly. In 1990 alone following the eruption in the mouse population after the 1989 rains, I retrieved over thirty large venomous snakes from houses and yards around Roxby and released them into the scrub. In typical years I catch less than ten.

Spring months following heavy rains see native pines browning off before releasing smoke-like clouds of pollen. This pollen and that of wattles that flower simultaneously gum up my eyes and those of other hayfever-suffering Roxbyites, keeping a rotation of hankies wet for at least a month. Grevillias, which have masqueraded as mulgas for months or years, show their true colours with impressive white blooms that excite the wasp, beetle and butterfly populations just before Christmas. The last of these waves of insects may rise to prominence over a year after the rain that fuelled the whole amazing cycle.

It is little wonder that the weather is the perennial conversation topic out here.

2
Lessons from the trees

'When we got to Woomera and saw that Roxby was another ninety kilometres, we wondered how you could live in such a desert!' exclaimed my mates from Adelaide. The site of Woomera was deliberately selected in the 1950s to be, as city slickers refer to it, in the middle of nowhere. Although I cringe at the derogatory terms for outback localities I have to concede that at first glance the horizon here is uninterrupted by anything above knee height.

Half an hour later my mates were surprised and relieved to be driving through an open woodland of native pines, mulgas and majestic western myall trees. 'You're bloody lucky that you've got all these trees', they continued, looking around at the Roxby woodland.

Fortunate definitely, but not lucky. The presence of these trees is not determined by luck, but by a whole series of factors, including historic rainfall events that would have been heralded by swifts and followed by swarms of insects. But the most important factor in structuring local vegetation communities is the soil. The flat gibber plains around Woomera and the other stony deserts of the outback seldom support trees. This is because water quickly runs off the clay soils that dry rapidly after rain. By contrast, deep sand can store water like a sponge and allow larger and longer-lived plants to establish. That is why there are trees in the dunes around Roxby but not on the flats at Woomera.

It is not only sand dunes that store water. Certain porous rock types, particularly limestone, store pockets of water that are sometimes accessible to the searching roots of trees. Sub-surface limestone is often indicated by patches of particularly lush-looking bullock bushes or paradoxically named dead finish trees that tap into water stored in cavities. Sometimes these cavities collapse into sinkholes or dolines that act as local sinks for run-off. The number of dolines lurking just under the surface becomes evident when we try to construct water storages. Sunday Hole and Monday Hole 'dams' on Roxby Downs Station are now gaping holes that swallow any water draining in from the catchment area. Likewise the water in several sewer ponds and stormwater dams around the Olympic Dam mine has suddenly disappeared when a doline has opened up underneath.

I was nearly a first-hand witness at a spectacular doline rupture. On a hot summer's day I was lured in for a swim after rounding up some kangaroos which had been trapped inside the desalination plant. This reverse-osmosis 'water-purifier' desalinates bore water from the Great Artesian Basin for the Roxby town water supply. The recently filled water storage pond was cool and clean and I splashed around with a few grey teal. A few hours later the entire contents of the dam, equivalent to about twenty Olympic swimming pools, disappeared down a gaping hole in a matter of hours. The engineers reckoned that it was sucked down at a rate of twenty cubic metres a second. Had it ruptured hours earlier I would have been swirled down the massive drain hole. From then on I was particularly aware of dense bullock bushes that could indicate underlying limestone cavities, and was adamant that planning engineers should inspect potential water storage sites for these tell-tale signs.

※ ※ ※

When I first arrived at Roxby the variability in vegetation on different dunes and inter-dunal swales confused me. It was not uncommon to have consecutive dunes vegetated almost exclusively by pines, mulgas, hopbush or canegrass. When coming over a sand dune it was almost impossible to predict whether the next flat would be a grassy swathe, a western myall woodland or a shrubland of saltbush and bluebush. What was the secret to this variability? Surely adjacent swales or dunes should experience a similar climate and have similar soils.

Over the years this mystery has slowly unfolded. While the soils with varying levels of limestone, sand, clay and gibber stone play a role, assaults on the vegetation over several centuries have also shaped the appearance of the region today. Grass growth following big rains can fuel fires, particularly in sandy areas, and these patchy fires will kill most of the trees and shrubs in their path. Unlike eucalypt woodlands that regenerate rapidly after fire, most desert woodlands die when burnt and may be replaced for decades by grass or shrubland. Violent storms may screw the tops off mulgas and hail may strip the branches from trees and shrubs in broad yet distinct swathes. In the space of a few minutes the composition of centuries-old vegetation may be obliterated in one region, but remain untouched 100 metres away. Dense low-lying stands of trees or shrubs may be killed during floods in areas that have been favoured by enhanced water accumulation for decades. In contrast, trees and shrubs may die in some patches due to drought but survive in other areas that have been blessed by a thunderstorm. Droughts, differential seed set, patchy pollinator distributions, localised stock, kangaroo and rabbit grazing all play a role. It is little wonder that most swales or dunes look different from the next.

Factors that I did not imagine would have a major affect on the distribution of long-lived trees were disease and parasites. In some wetter environments, and particularly those exposed to disturbance or introduced organisms, large-scale tree deaths can occur rapidly. Huge swathes of massive jarrah trees and other natives died following the spread of the *Phythophera* fungus through wetter areas of Australia. Equally devastating were the deaths of the mighty chestnut trees of the eastern United States, all but wiped out by an Asian fungus called chestnut blight. In the same areas a beetle-transmitted fungus from Europe called Dutch elm disease has killed millions of native elm trees. But these were not tough desert trees and the fungus and beetles that afflicted them were not a problem in the undisturbed natural environment. The hardy trees around Roxby were tough enough to have withstood all of the trials of several centuries of hazards, during which they surely must have experienced just about every possible climatic or biological insult. Furthermore, they were largely isolated from introduced species that could expose them to new diseases or parasites.

Of all of the local trees the western myall is perhaps the most rugged and longest lived. Western myalls are majestic eight-metre high trees that could be represented by broccoli stalks in a scaled-down model of the landscape. Some are nearly a thousand years old and they are, without exception, magnificent examples of persistence in a harsh environment. Therefore the die-off of myalls at Roxby which was first noticed at the beginning of 1999 was met with surprise as much as shock. At first we suspected that the drought combined with the stress associated with earthworks or dust might have pushed these splendid old specimens over the edge. However myalls were also dying at Glendambo, over 100 kilometres to the west. Thousands of myalls were dying in remote areas far removed from towns, roads or other recent influences that might have been implicated in their demise. Little black scales found on the leaves of all of the dying trees were thought to be responsible and the experts were called in.

Jon Martins is a little English entomologist who is living proof that some scientists resemble their study creatures. At a glance it was clearly evident that Jon did not wrestle crocodiles or track tigers through jungles. His speciality was white flies, which I learned were not actually flies but tiny little white-winged bugs whose larvae sucked the juices out of plants. It was the larvae that lived in the black scales that we could see on the leaves. Jon was as excited as a white-fly buff could be. At Roxby he witnessed the largest population of white flies that he had ever experienced and, what's more, they did not have a name. Jon could tell us that it was an undescribed species of *Zaphanera* that had never been collected before. The 'fly' that was killing our favourite trees was not known to science although there were literally billions of them sucking on local myall trees. Jon found several other species of undescribed white flies around Roxby and lamented the poor state of our knowledge of this apparently fascinating group of insects. The Roxby community was not so fascinated. We were more concerned about the future of our myalls and we wanted to know how to control these pesky 'flies'.

Jon passed the buck to a large bald bloke who turned out to be fortuitous proof that not all scientists resemble their life interests. The assembled Roxby Downs Field Naturalist Club turned our attention to another entomologist who had come to help us out. Andy Austin, from Adelaide University, was the Australian guru on parasitic wasps. Andy had found one of these little wasps feeding on the white flies at Roxby. Like the flies, this wasp was new to science. Andy confirmed that 'his' wasps were killing some of 'Jon's flies' which were killing 'our' myalls.

'They are obviously not doing a very good job!' exclaimed Trevor Porter from the golf club which had suffered a severe loss of myalls on its fairways.

'What is limiting the wasp population levels so that they can not control white-fly numbers?' we asked.

Andy had nowhere to turn. Because the wasp was an unknown unstudied species, he obviously could not say whether the wasp was lacking other key resources. Our poor knowledge of the functioning of arid ecosystems had been pathetically exposed and we were left with no option but to monitor, brainstorm and research a range of potential control techniques. Fortunately, within a couple of years the wasp populations increased and brought the white flies under control but not before hundreds of groves of myalls were killed.

While freak episodes such as the white-fly infestation, floods, fires and storms can have a devastating effect on the local vegetation, there is something else that has an even greater influence. Despite widespread groves of adult trees, young trees of any species are alarmingly scarce. Sir Douglas Mawson was concerned about the lack of regeneration of mulgas and other trees in the arid zone way back in the 1940s. Another wise old man of the bush, George Bell, harbours similar concerns. George has never seen a young mulga tree in his sixty years of work on Dulkaninna Station. Although mulgas are not as abundant on the Birdsville Track as they are elsewhere in the outback, they are nevertheless widespread on Dulkaninna. Many of the station fences built over a century ago used mulgas, which attests to both their durability and abundance. Given that young mulgas less than thirty years old are clearly recognisable from older trees, George's observations, or lack thereof, suggest that there has been negligible recruitment of mulgas in his country for at least 100 years. To explain this quandary, we can look beyond myalls and mulgas to the tallest tree in the Roxby region.

※ ※ ※

Old Max Thomas, an Aboriginal man who lived on Andamooka Station, told me that 'coorly' was the Kokatha word for the native pine, an important and characteristic tree in this country. Coorlay Lagoon, just south of Roxby, was named (yet misspelled) after the pines that grow in the sand dunes around its shores. In 1989 the 'lagoon of pines' revealed some incredible facts about one of Max's favourite trees. Several large old gnarled pines were inundated by the flood and died as a result.

The death of these pines taught me two lessons. Firstly, pines don't like getting wet feet. Secondly, the 1989 rains must have produced the largest flood since before those pines germinated, several hundred years ago. Near the base of some of the pines was an army of smaller dead pines, about the size of large Christmas trees. These smaller pines were all about the same height and had apparently germinated at the same time. Counts of the growth rings indicated that these recently drowned smaller pines probably germinated during the big rains of 1946.

It was not the fact that all of the Coorlay coorlys had drowned that amazed me most. It was the realisation that only once in over 200 years since the older pines were established had any youngsters germinated. I started looking around the district to see if I could find pines that germinated at other times. Despite pines being one of the most common trees around I could only find eighteen groves of youngsters. All of these neat stands of young pines had germinated in either 1946 or 1974. Given that these years were incredibly wet, why weren't the young pines crowding out all of the sand dunes, like the closely related native pine of the Flinders Ranges has been doing since the wet mid-1970s? Roxby's pines definitely had no problems germinating, for in 1989 thousands of seedlings were poking their way out of the damp soil within months of the rain.

I suspected that the answer lay with rabbits or domestic stock knocking off the seedlings before they had a chance to establish. Therefore, with the aid of a few helpers I placed tree-guards around a potential forest of young pines within a region that was free of sheep and cattle, and left other seedlings exposed to rabbits. Within a few months every tagged and measured seedling had died. Not one succumbed to rabbit grazing. Every tree inside and outside the tree-guards simply dried up and died. I couldn't believe it! Earlier that year we had the biggest flood on record, yet after a couple of dry months the seedlings were dying.

A few years later the lagoon of pines revealed another interesting secret. Perched at the very summit of the island where the really old and forty-five-year-old pines had drowned was another cohort of little pines. They had germinated in 1989 in the few square metres that had escaped inundation and hence were given the perfect start to life.

With no competition from other pines and, most importantly, a ready supply of fresh water for several years, the young pines continued to grow while the others in the bush died. A few other young pines also survived along the edges of roads or adjacent to unguttered rooves where water concentrated. One grew next to a dripper in the newly built Roxby Downs cemetery, which had still not been 'occupied' a decade later. The only places that the seedling pines had survived were where they were given assistance in combating the three metres of annual evaporation, which is nearly twenty times the average rainfall.

On a series of sand dunes just north of the Roxby Downs homestead where one of the groves of 1974 pines was found, there were also mulgas and myalls of similar age. Being close to the homestead and water this site would have been grazed by stock ever since these trees germinated. The logical explanation for this area of rejuvenation was that following the big 1974 rains this region received localised rainfall from one or more storms. While seedlings in surrounding regions perished through water stress, the rain experienced here was enough to keep the young trees growing to the stage where they could endure the next dry spell. By studying the groves of young pines I established that seedlings need two years of above average rainfall in order to recruit. The scarcity of young trees around Roxby, Dulkaninna and other regions was largely because the rainfall for most of the past century has been insufficient to allow recruitment, except where fluky rains overlapped for several consecutive years.

This rare, unpredictable and patchy rainfall that is necessary for tree recruitment influences more than just the trees. Pines and myalls, like the gums and gidgees further north, were used extensively for fence posts. A close inspection of many mature trees throughout the pastoral regions often reveals signs of previous cutting. Fortunately, unlike most pines which have a single trunk, the native pines in the Roxby region are multi-stemmed, or mallee form, and hence will often regrow after the main trunks have been removed. In addition to providing tough termite-resistant timber for fences, yards and even the construction of the original homesteads and huts, the local trees provide essential nesting sites for many birds. At some stage in the future the gnarled old trees will fall over and the bluebonnet and mulga parrots will have to look elsewhere for hollows. Eagles and kites may be left without trees to construct their nests in.

Now when I drive into the Roxby dunefields I look at the groves of pines and myalls with awe. They are postcards from the past, living fossils that germinated centuries ago when the weather patterns were different from those experienced since Europeans first laid foot in the outback. They, along with the parrots and birds of prey that use them, are the survivors of centuries of challenges by fire, flood, drought, axe, chainsaw and grazing. Who knows how long these dinosaurs can survive? Will the climate produce another run of exceptionally wet seasons that will permit the recruitment of another swathe of trees that may persist halfway through the current millennium? Or will the pines and myalls, like the extinct forest plants which can be seen in local fossil deposits, be relegated to history?

3
Every good drought is broken by rain

'I can't understand all of this raving about the rain', said the laconic bloke propped against the workshed wall. 'All it brings is bloody flies and stink bugs and a "fungus" all over the place'. Adam continued his diatribe without pausing for breath. The fungus that he was referring to was the flush of ephemeral growth that partly concealed the red soil and gibber stones that are typical of much of the north. For tourist operators such as Adam Plate from Oodnadatta, rains bring mixed blessings. His ramble also admonished the cut-up roads, the mosquitoes and the mud. Although full lakes and wildflowers are a real magnet for tourists, travelling and camping in the outback is generally easier and more comfortable in the dry.

Adam's somewhat tongue-in-cheek haranguing is mirrored by some environmental issues. While the big rains are obviously integral in shaping the vegetation and fauna communities of the north, rain also places the relatively unadulterated arid zone of Australia under stress. When it rains heavily the hardy drought-tolerant plants and animals must contend with rivals or predators from wetter areas.

To understand the role of droughts in structuring and maintaining local plant communities I was given a few clues by Frank Badman, 'The Botanist' at Roxby. Although not known for his electric wit, whenever we had a big rain Badman used to remark 'Every good drought is broken by rain.' When he started conducting formal research on plant distributions his dry rhetorical comment took on a whole new meaning. Frank collected an enormous amount of information that revealed a relationship between rainfall and the distribution of weeds that made logical sense. Nearly one third of the plants that Frank surveyed near Port Augusta were weeds, or 'aliens' as he described them. As he headed north this percentage dropped off incrementally, with less than seven percent of the plants around Roxby considered aliens. Weeds like Wards weed, onion weed and a range of thistles generally don't make it too far north. Occasionally we see Salvation Jane at Roxby or three-corner jacks at the lakes near Woomera, but they quickly die out in dry times. Noxious pest plants like Bathurst burr and horehound persist at a few dams but have not spread through the countryside like they have down south. The southern pastoral region is more exposed to weed seed sources from gardens, stock feed and fertilised agricultural regions. It has also been exposed to these weeds for longer than the north of the State. But it is the lack of rain that is the most important reason for the scarcity of weeds further north. The long droughts help keep out weeds and make inland Australia the special and relatively untouched place that it is.

Drought conjures up images of desolation and desperation. For humans and domestic animals, long dry spells are devastating. Images of animals perishing are distressing but droughts are as important to desert ecosystems as rain. I was probably getting a bit carried away when I wrote earlier that 'rain was everything'. Droughts are like Brussels sprouts: we don't particularly enjoy them but deep down we know they are good for us.

It is not only the plant weeds that can move in after rains. In wet years animal 'weeds' invade the outback and would persist if the rain was more regular. Large flocks of starlings, including the 500 that I saw at Dulkaninna Station in 1989, sometimes turn up. Other introduced bird species along with house mice, introduced rats and insects also occasionally establish populations in the arid zone. While the mice persist in low numbers most of the others get driven back in dry years.

Dry periods also limit the spread of certain native species, thus allowing desert-adapted species to thrive with minimal competition. Yes, drought is actually responsible for increasing and maintaining the diversity of species found in the outback. For instance, sleepy lizards are not cut-out for really dry conditions, which in part explains why they are not found north of the Cooper or William Creek. However, a suite of other lizards, including gibber dragons and bronzebacks, are restricted to these dry deserts. Native pines and western myalls also cut-out just north of Roxby. A whole host of bird species reach their northern limit as you head north from Port Augusta towards Lake Eyre. They make way for the true desert specialists including the gibber bird and several species of grasswrens. While western grey kangaroos only spread north into the dry territory of the red kangaroo following a good run of seasons, they die out or move south when times get tough.

Even more difficult to comprehend than the deaths of normally hardy species like sleepies and kangaroos, is the drought-induced death of the robust plants that define the outback. Several months ago the saltbushes had shed their leaves in a last-ditch effort to make it through to the next rain. Many did not make it, their bare stems gradually turning brown and brittle in the relentless sun. A few of the saltbushes had survived, especially those growing in slightly deeper sand or in small patches that received a smidgin of extra water that ran off bare rocky patches in the last drizzle. Their leafless branches retained a tinge of green and a slight spring in stark comparison with the rest of the crackling grey–brown twigs. Like the sleepies and the *Sida*, the progeny of survivors living in slightly wetter areas would eventually replace the saltbush that succumbed to the drought.

Droughts have also limited the spread of agriculture, undoubtedly saving many desert species. Pastoralism utilises the natural vegetation in a largely unaltered state, thus at least theoretically allowing natural ecological systems to keep functioning. Farming or cropping removes the native vegetation and it is not only the native plants and animals that suffer. The loss of soil from dry agricultural areas is particularly evident in the sad remnants of mallee remaining near Port Germein. These clumps of trees are perched on islands of soil on the side of Highway 1, up to half a metre above the eroded agricultural desert. The topsoil from the farmed paddocks was the key ingredient to productivity in this marginal area and was blown off after decades of dust storms. Even if they are never farmed again, it is unlikely that these paddocks would ever be able to support the equivalent productivity of plants and animals of the adjacent, uncleared areas.

Understandably, we tend to have a mammalian bias of what is 'good' or 'bad' country, and what is a 'good' or 'bad' year. Many insects and reptiles may think otherwise. Ask any eight-year-old boy what type of weather is the best for catching lizards and he will tell you that it is sunny days, of course. Most lizards need to bask in the sun to warm up in order to hunt, digest food or breed. Rainy days, and particularly long cool cloudy periods, can disrupt a lizard's schedule. Very wet seasons where it rains nearly every week for a few months on end can seriously limit the feeding time for lizards. In turn they may lay less eggs and the juveniles may struggle to beef up before the winter sets in.

Not only can wet years cause hardships for some lizards but it seems their prey can suffer as well. Many lizards, including the most abundant local skinks, feed predominantly on termites. Two of the most common local geckos, the beaked and fat-tailed geckos, feed on nothing else. The numbers of these geckos and possibly their reproductive output appear to be affected by the abundance of termites. Since both geckos and skinks that specialise on termites typically produce many eggs even during dry summers, termites appear to be a reliable food source regardless of rainfall. Therefore, many termites, geckos and skinks typically do not suffer in years that we consider to be 'bad'.

It is not only the lack of rain that helps promote biodiversity, it is also the boom or bust nature of it. Many trees like the pines that germinated in 1946 and 1974 require an exceptional series of rains to ensure their survival. If we received a relatively regular 150–200 millimetres of rain each year many of our plants and wildlife would not be able to reproduce. 'Spreading' our rainfall of about 170 millimetres a year into boom and bust years is vital for the recruitment and survival of many species that occur in the outback. But there is more to the timing of rainfall than that. Roxby Downs is located approximately where the summer monsoonal rains and winter cold fronts exert equivalent influence on the rainfall. We are just as unlikely to receive rainfall in any month. In some years many of our wildflowers or trees don't flower at all. When they do flower, the type of flowers that bloom are dependent upon the timing of earlier rains. So, unlike places with more predictable rainfall we can not host wattle, daffodil or rose festivals. The only exception is the Arid Lands Botanic Garden at Port Augusta, 260 kilometres south of Roxby, which hosts a spectacular *Eremophila* festival in spring. The bewildering array of these desert-specialist plants is irrigated to ensure that they flower with gusto on cue. In the bush, without the benefit of a regular water supply the variable timing of rains has a dramatic influence on the types of plants that grow and hence the types of insects and birds that arrive.

Summer rains stimulate germination in grasses, which provide a boon for grasshoppers and seed-eating birds like budgies, cockatiels and pigeons. Winter rains generally produce more wildflowers, encouraging caterpillars, chats and woodswallows to proliferate. But even this seasonal breakdown simplifies the response of plants to rainfall enormously. Depending on the timing of rains the dunes may be cloaked in green with grasses, purple with the flowers of parakeelya, white with poached egg daisy and wild carrot, or yellow with crotalaria, everlastings and groundsel daisies. After some rains the plains are seas of red Sturt's Desert peas, orange 'egg and bacon' peas, or purple Swainsonas. Even the type of grasses that germinate depend on the timing of the rain.

Climate changes have seen waves of different species come and go as evidenced by the fossils of plants and animals associated with rainforest or inland seas which are still being found locally. Similarly, we can expect many more changes as the climatic balance shifts. Where climate change is accelerated or driven by the exhaust fumes of modern society, we have a responsibility to predict and prevent such influences wherever possible. If we don't, the outback will be a very different place in the future.

4
Don't bother with the forecast

We have established that big rains and big droughts are as much a part of the outback as blue skies and flies. We also know that the timing and intensity of these climatic statements allow experienced observers to predict most of the responses of plants and animals. What is much harder to establish is what the forecast will be, despite the yearning of pastoralists, tourists and ecologists for exactly this information.

Imagine you are presented with a river red gum seedling. You can plant it anywhere you choose and your budget allows you to take any drastic measures to ensure that in 100 years the tree is identical to the one in your favourite Hans Heysen painting. Picture the gnarled old trunk, the broken limbs, the holes and boles and mistletoes. After a century of manicuring I suspect that your great great grandchildren would look at the result and start inquiring about how difficult it is to 'doctor' an old painting to match your tree. Even though we have the technical knowledge to predict whether the tree will be tall or short, vigorous or stunted, heavily or sparsely foliated, the shape of any gum tree is subject to such a huge range of factors that its shape will never be recreated.

In recent years meteorologists, oceanographers and climate historians have started to predict weather conditions for months and even years into the future. Some long-range forecasters use sunspot activity or the alignment of the planets, while the most promising trends have come from the analysis of ocean temperatures. These new scientific and pseudo-scientific approaches are based on reasonable assumptions and measurable indices. However, as far as predicting the rainfall at a particular location in outback Australia, they are sadly inadequate. Long-range forecasts are no more likely to be correct than predicting the shape of a gum tree. To be brutally honest, long-range forecasts are crap.

Ocean temperature fluctuations drive hemisphere-scale atmospheric moisture patterns that have been given names like Spanish dancing horses. El Niño and La Niña strongly influence climatic patterns in some Southern Hemisphere coastal regions including the eastern seaboard of Australia. It is no coincidence that this is also where most of Australia's meteorologists, agriculturalists and ecologists live and hence El Niño and La Niña have recently assumed cult status. However, the further inland they get the more distracted these horses are and the more erratic their influence becomes.

Most of the big rains in outback Australia—the lake fillers—originate from cyclonic depressions to the north-west of the continent. Ocean temperatures influence the frequency of cyclones but not their paths. Predicting the course and severity of a cyclone while it is doing its stuff is difficult enough. Remember Cyclone Steve in early 2000? It first wreaked havoc near Cairns on the north-eastern coast of Queensland then crossed the Gulf of Carpentaria and Kimberly regions before petering out near Esperance on the Great Australian Bight. In two weeks Steve completed more than a half circuit of Australia, typically dumping 100–400 millimetres of rain in its path. Swathes of rain-bearing cloud swirled inland from several directions during Steve's exploits. Forecasting where in the outback the resultant moisture will travel, months before cyclones have even formed, is impossible. It is like predicting the path of a spinning coin on a table.

According to the particular forecaster that WMC throws its money at, 1999 was supposed to be an extremely wet year at Roxby. The ocean currents and timing of atmospheric troughs suggested that the weather would mimic that of 1974. We were told to expect one of the wettest years for decades. We got nothing. To be precise, Olympic Dam got just sixty-nine millimetres, about a third of our average, while just down the road at Andamooka they got 100 millimetres, less than a third of their 1974 recording.

But the forecasters were not totally wrong. Lambina Station near Oodnadatta received 265 millimetres for the year and the country looked more like a rice-paddy than a desert. Fine-tuning of weather forecasts below the scale of several hundred kilometres is apparently not possible and hence these forecasts are of no value to land managers. Even in the wettest of wet years some parts of the inland receive less than their average rain. The opposite also holds true for drought years, with some areas defying the influence of the Spanish dancing horses.

I am convinced that the meteorologists with the barometers and ocean probes can only make semi-informed guesses, but how about the old-timers on the land? Our pioneers obviously had no idea how much rain to expect. Until the government surveyor Goyder mapped the ten-inch rainfall isohyet, north of which he suggested that cropping was not viable, pioneering farmers held the misconception that they could grow wheat as far north as Farina, just across Lake Torrens from Roxby. Even more bizarre was the assertion that rain would 'follow the plough' and a scattering of ghost towns attests to their mistake. The location of several early homesteads such as Parakilya, Roxby Downs and Purple Downs, which were all built on the edge of swamps, also showed the lack of understanding of the pioneers when they were subsequently inundated. Most if not all of the early farmers misjudged the local climate. But what have we learned about forecasting the weather since then?

My personal experience suggests that it is likely to be windy for the Marree races in June as well as the Birdsville races in September, and that rain is likely just before the Oodnadatta or William Creek races in April or May. Also it is always hot and dry enough for the fire truck to douse revellers at the Glendambo B & S ball recovery in late November. Apart from these few dates each year I am no better than the crystal ball meteorologists who watch the Spanish horses. However, several local identities have more clues.

Bobby Hunter relays a story told to him by the blackfellas from Oodnadatta. Apparently cadneys, which herpetologists know as bearded dragons, perch on posts or bushes with their mouths open when it is going to rain. These thirsty lizards are anticipating a drink and since they are such great survivors in the outback they must know what is going on!

Emus usually nest in winter, at the same time as their astronomical equivalent—a dark silhouette in the Milky Way—appears just above 'the dingo' on cold clear nights. The male emu typically sits on a nest on the flats, sometimes partially secluded behind a low bush. Old Max Thomas, the Aboriginal bloke who liked pine trees, told Leo McCormack from Roxby Downs Station that on the rare occasions when emus nest on the sand dunes it is a sign that there is going to be a big rain. Although Leo once found an emu nest on a dune before rain I remain to be convinced that a bird that can not run and think at the same time has the mental prowess to predict floods months in advance.

Kevvy and Shane Oldfield swear that their frogs at Clayton Station on the Birdsville Track make a hell of a racket a day or two before rain. The Oldfields would sooner turn vegetarian than imitate a frog call so I'm not entirely sure which frog species they are referring to. However, I would put my money on the little brown tree frogs rather than a burrowing frog being their rain indicators. It is not surprising that frogs will respond to a reduction in air pressure that often accompanies rain. People who keep frogs notice that they start calling like noisy green barometers when a change is arriving. Frogs also get carried away when the air pressure declines on a plane flight. Several times I have collected frogs from remote localities for a museum or university researcher and concealed them under my shirt or in my hand luggage on a flight. Soon after take off the frogs that may have been quiet for several days start calling. Females may even lay eggs in anticipation of the deluge that normally follows the drop in air pressure. It is hard keeping a straight face when other passengers start asking each other if they can hear frogs, and contemplating whose bag they might have stowed away in.

Unfortunately frogs don't always get it right. At the next station up the track from Clayton Station, Daryl Bell has been caught out several times by frogs calling around the Dulkaninna Station homestead when no rain has eventuated. Daryl reckons that spider webs draped over bluebushes, or mulga trees coming into bud during a dry spell are more reliable indicators of impending inclemencies. Perhaps the spiders become more active because moths and other flying insects build up before the rains. Phil Gee from Coober Pedy, who has spent more time around a campfire than most other bushies, predicts rain within a day or so of having his campfire bombarded by flying insects.

Many people believe that a red sunrise or a ring around the moon means that rain is coming, but this is easier to say than to prove. I have still not met anyone who is prepared to bet on such predictions. Most bushies agree with Cactus Williams from Lambina Station that water-birds start turning up and ant activity increases just before big rains. Ants may march out of creeks and start climbing up walls and trees before the first big drops splatter onto the dusty soil. Cactus also maintains that ducks and black 'shit chooks' call as they fly over on their nocturnal pilgrimages before rain. Like arthritic elbows and croaking frogs, it is likely that they are responding to changes in atmospheric pressure that may or may not result in rain. Ducks, ants and moths sometimes get it wrong but we tend to quickly forget these *faux pas* if it doesn't rain. They are at best a guide that rain may be on its way within a few days.

So I remain unconvinced that any outback animals or identities can provide a reliable early indication of big local rains. If you think I am being harsh try a little experiment. See if you can find any meteorologist, cloud gazer, emu watcher or old-time legend who can predict the rainfall for a particular inland location for the coming year. There is little value in forecasting rain for fifty or 100 kilometres away. If they are any good they should be able to tell you which seasons will be 'wet' or 'dry' at their selected rain gauge. Being generous I would accept as being accurate a figure between 50–200% of their prediction for each three-month period. To demonstrate that any 'correct' predictions were not a fluke get your forecaster to predict the seasonal falls a year in advance for three consecutive years. I am confident that few experts would take up the challenge and none would be successful. Despite the pride and confidence we have in our intelligence and technology, long-range weather forecasting in central Australia battles with the same unpredictable processes that determine the shape of an old gum tree.

No matter what the forecast is, or how the plants and animals are responding, outback residents still glance at the north-western sky every morning and night. Big black clouds have got to be a better indication of rain than long-range forecasts, ants or frogs. Nowadays many pastoralists are turning their backs on the erratic signals from nature and instead are relying more and more on satellite images transmitted through the Internet. However, like the sunrises, frogs, insects, birds and maybe the lizards, satellite images only give us a day or two's notice of impending rain. In order to plan optimal timing for stock movements, vermin control, outback functions, road maintenance, dam cleaning and most other activities, we would ideally like to know weeks or months in advance. Like our plants and wildlife, we must respond to the rainfall rather than trying to forecast it, and this has huge implications for pastoralists, wildlife managers, tourist operators and, of course, native plants and animals. Maybe the uncertainty is a good thing, like not knowing when the siren is going to sound in a close footy game or not being able to predict the shape of a gum tree. Maybe this unpredictability shapes the unique bush psyche. Maybe we should stop trying to predict the rain.

PART TWO
Just add water

5
Refugia

The unchanging rhythm of the train provided an appropriate soundtrack for the monotonous view outside. Gone were the green hills, kangaroos and emus that I counted on family trips. For hours I had seen nothing but flat dry ground, red rocks and half-dead shrubs. 'Have a look out of the window,' Mum said. 'There's Lake Eyre.' To a six-year-old lad on the *Ghan* the lake was big, white and boring. I couldn't wait to get to my cousins' place in Alice and go to see the Rock. I did not see the lake for another twelve years.

Since I first saw Lake Eyre it had filled in both 1974 and again in 1984, but had subsequently reverted to the dry saltpan that I vaguely remembered. Now the lake was my entire world. I had been woken before dawn and driven to Gosse Springs, a scatter of low green mounds of sedge on the shore of Lake Eyre South. The blokes who dropped me at Gosse injected a red dye into the outflow of a couple of the springs. They had then quickly shown me how to dye-gauge spring flow and had disappeared off to other springs. As the morning sun was beginning to burn the crest of Hermit Hill I filled a numbered glass vial with spring water downstream from the dye-injection point. For the remainder of the day I was to collect water samples at designated locations along the outflow tail of the springs. At first the samples were taken every few seconds but as the dye passed down the tail sampling intervals spaced out to five, ten then every sixty minutes.

I was on my University holiday being paid to do thirty seconds work each hour in one of the most bizarre landscapes on the planet. WMC were monitoring the flow rate of the springs that are natural outflows of the Great Artesian Basin. The GAB, as it is widely known, is an enormous series of water-holding rock beds that underlie 1,700,000 square kilometres of central and north-eastern Australia. Water enters this basin through porous rocks along the Great Dividing and MacDonnell ranges and slowly filters thousands of kilometres south towards Lake Eyre over a million or so years. This water eventually seeps upwards through semi-porous rocks and soil to evaporate on the surface or flows more quickly up fractures to form springs. These mound springs, so named because some develop mounds of wind-blown sediment or precipitated limestone, are important heritage sites supporting rare plants and animals. The mining company was obliged to ensure that their water extraction from the GAB had minimal impact on the springs.

With the exception of the grazed stubble of *Cyperus* sedge that rimmed this waterhole I could not see another plant on the horizon. This was the haunt of the famed Lake Eyre dragon, a tough little lizard that lives in one of the world's most inhospitable environments. I revelled in racing out in a certain direction for twenty minutes, looking for dragons while I regained my breath and then jogging back to take more water samples. I was amazed at the animals including earless dragons, geckos, scorpions and even a nesting spotted nightjar I found in this formidable landscape. A couple of generations ago the story would have been much better.

As a six-year-old rumbling past the lake on the *Ghan* there were still people who could remember seeing the pademelon, or desert rat-kangaroo, *Caloprymnus campestris*. This small kangaroo was just larger than a rabbit and lived on the stony plains near the Birdsville Track, just to the east of Lake Eyre. *Caloprymnus* lived in shallow scrapes under sparse low bushes, sometimes lining their abodes with grass, leaves and sticks. The last specimen was collected in 1936, but the old-timers from the Track reportedly last saw the species in the 1950s. *Caloprymnus* is now officially regarded as presumed extinct. That means gone ... forever ... probably.

When the ominous 'presumed extinct' status is conferred on a species it invariably indicates that European settlement has been responsible for their demise. 'Presumed extinct' means that the species has been waiting in the departure lounge for definite extinction for less than the prescribed fifty-year period before a line can be drawn through its name, and its future. 'Presumed extinct' has a hollow resonance when it follows the scant description of a species known only to a handful of white observers, but which was presumably an integral part of the fabric of life for hundreds of generations of Aboriginal people.

However, several tantalising reports from the 1970s and 1980s gave biologists hope that *Caloprymnus* might be hanging on, somewhere. If the little desert rat-kangaroo did survive I wanted to find it. What greater thrill would there be than to rediscover a species that was presumed extinct and then to assist with its recovery? I was inspired by stories of golden lion tamarins, bridled nailtail wallabies and birds from Lord Howe Island that had, through clever captive breeding programs and reintroductions, been rescued from seemingly inevitable extinction. However, no-one could do anything for *Caloprymnus* or any of the other presumed extinct species unless some were found.

In early 1990 while I was contemplating the fate of *Caloprymnus* and several other mammals that had apparently disappeared from central Australia, a scientific paper was written by Steve Morton, arguably Australia's pre-eminent arid zone ecologist. From his CSIRO base in Alice Springs, Morton attempted to explain the worst contemporary mammal extinction rate in the world. He was working over theories that had been espoused for years since Hedley Herbert Finlayson first documented the atrocious plight of our desert mammals in the 1930s. Feral and introduced animals were still considered to be the primary cause of the declines in populations, although changes to burning regimes, drought and in some cases hunting were also thought to play a role.

Although most of the theories and explanations were not new, Morton emphasised the role of 'refugia'— the somewhat dry term for describing an oasis. The refugia that Morton alluded to are typically less glamorous than the palm-lined waterholes that have inspired the name *Oasis* for roadhouses and restaurants across the outback. Rather, refugia are areas where nutrients and moisture accumulate from adjacent regions to provide an asylum for desert-dwelling mammals during droughts. Unfortunately, as Morton pointed out, these richer patches were also attractive to stock, rabbits and other feral animals and when it came to a battle for survival the native species lost.

Morton's paper enthused me to search for wet or fertile areas in the desert that had escaped degradation by introduced species. Mound springs were an obvious candidate. These natural wetlands vary in size from small seeps, like some of the springs I had monitored at Gosse, to fast-flowing geysers that overflow their vents to form extensive wetlands. It was these larger springs that I hoped might provide the last refuge for some of the mammals presumed to be extinct.

Unfortunately all large mound springs are not all that they should be. Fred Spring near Lake Eyre South had been described to me as 'the arsehole of the world'. I had parked at Fred Spring, which had an old pastoralist's bore punched through it, to measure its flow with David 'King George' Whiting. King George, an environmental scientist who worked with me, had the job of measuring spring flows and bore pressures around the WMC Borefield. I was compiling an inventory of bird use of different springs. Neither of us achieved our objective that afternoon.

A cow had become bogged in the bottomless morass of the spring. Only the top half of her back and her head were exposed above the pugged brown slop. A dingo that was patiently waiting about 100 metres away had already chewed her ears off. We couldn't leave her there in such a state so we left our gauges and binoculars in the car and walked over to help.

Springs can be deceptive. On the edges of the outflow, or 'tail', the wet soil is typically compact. However in the 'vent' where the water percolates up from underground the mud has the consistency of chocolate mousse. This cow looked as if she had been dropped straight into the vent from a great height. We could not get our car close enough to her to drag her out with a rope and we did not have a gun to put her out of her misery. With no other alternative we stripped to our jocks and jumped in with her.

The cold stagnant mud stung as we immediately slipped in above our knees. Our bare feet nearly froze in the cold June mud and scraped against the bones of cattle that had previously succumbed in this quagmire. I made the mistake of going towards her arse end and it was immediately apparent that she had been incarcerated for several days. Although she had not been eating, her digestive system had been busy and her flailing hind legs had churned litres of green liquid into the stinking broth. Fortunately I did not have to look at her pitiful eyes.

By pulling on her tail that was at about chest level I was able to free one of her hind legs; the other leg painfully raked against my bare feet a metre under the surface. We were able to get a rope around her leg by plunging it through the sloppy green broth. By pulling on the rope and pushing her head and forequarters we managed to roll her onto her back. Over an hour later after an exhausting series of rolling, dragging and pushing manoeuvres, we managed to extricate her out of the bog and onto firm ground. To our dismay she lay heavy-headed and motionless. I figured that her legs had probably cramped-up so I tried bending and massaging them, aware that at any time she might regain their use and send me flying back into the spring.

But there were no flailing kicks or defiant bellows. About ten minutes after we got her out she took her last breath. We were absolutely buggered, angry and disappointed. The onlooking dingo cocked his head as if to ask, 'Why did you bother?' All but our heads were plastered in foul-smelling, green–brown sludge. Blood oozed through the mud on our feet. I looked at the spring and decided that it would have to be the last place on earth that any rare mammals would survive. It was a death trap, not a refuge, even for the extinct *Diprotodon* whose fossils were found in springs near the Gammon Ranges.

King George and I wandered around on a nearby sandhill waiting for the mud to dry so that we could brush it off with a sand bath. To the north was the vast expanse of Lake Eyre South, perfectly reflecting the blue sky. The lake had filled as a result of the March 1989 rains and three black blobs floated in the middle of this silvery mirage. These islands were mound springs formed along a fault line that bisected the lake. Maybe these springs were different; there should not be any cattle out there to pug them into a stinking slop. Nor would there be any rabbits that mow the sedge on mainland springs like a bowling green. Maybe, just maybe, these mound spring islands might be the refugia that I was looking for.

I went home and did a bit of reading. Lake Eyre was formed in the mid-Miocene and most of the mound spring groups formed in the late Miocene. The islands in Lake Eyre South therefore probably emerged after the lake had formed. This made them even more interesting, as anything living out there must have been able to successfully colonise these islands from the mainland. Since Lake Eyre has shrunk over geological time the islands have probably been even more remote in the past than they are now. I remembered from biogeography lessons that remote islands often support unique types of plants and animals.

My mind wandered to an island utopia with permanent fresh water and protected by a huge moat of salty water, and I became even more convinced. There would not be any foxes or cats and the islands should be exactly the same as they had always been. Feral predator-free islands off the coast of Western Australia and South Australia have harboured several mammals that have recently become extinct on the mainland. Why shouldn't islands in the centre of Australia, in the heartland of extinction country, be even more important? Furthermore, the Lake Eyre South islands are incredibly isolated and difficult to access. I doubted whether many people had set foot on the islands, let alone had a decent scrounge around for rare critters. The best way of getting out to them would be by boat and the best time for that was when the lake was still nearly full. However, setting foot on the islands was destined to be even more of a challenge than I had imagined.

The following week King George and I had the opportunity of checking out the islands from the air. WMC had chartered a little four-seater plane to fly us to Jackboot Bore, between Lake Eyre North and South. The causeway between the two huge lakes was impassable and flying was the only way to access this important monitoring bore. Our boss, Barry Middleton, decided to come too, a flight over Lake Eyre in flood was too good an opportunity to miss!

From the air we could see *Phragmites,* the tall grass that grows in springs, on the two islands that were virtually in the centre of the lake. This reassured me that the springs that had formed the island mounds were still active. The other island was too far to the west to see clearly. The closest access to the unnamed centre islands appeared to be from the east, adjacent to the Gosse Springs that stood out as little green dots against the bleached white background. We incorrectly assumed that the closest access would be the easiest.

As we approached the short bush airstrip adjacent to Jackboot Bore a large green cross was clearly visible. While King George was taking his bore pressure readings I wandered off to inspect the cross more closely. It was formed by a luxuriant growth of turnip weed that was growing through the white bones of cattle that had perished at the bore and been dragged into this spooky formation. The cross turned out to be a bad omen.

While the plane lumbered up the short runway on take-off Barry's ears suddenly became redder. I knew there was a problem. He had pulled rank on us and jumped in the front seat, and my forward vision consisted entirely of his snow white hair contrasting vividly with his flushed anxiety gauges. Seconds later the plane started bumping violently. I could see from my window that we had overshot the airstrip but had not yet taken off. My anxiety gauges also started red-lighting. Suddenly it was over, the bumping stopped and I watched as a barbed wire fence whipped underneath the climbing plane. Three of us breathed a sigh of relief. Then the pilot, a local bloke named Peter, calmly uttered five words that a passenger never wants to hear, 'She doesn't want to fly.'

I looked out my window as the left wingtip sliced through a needlewood tree. Out of King George's window I saw 'his' wingtip hit the ground, breaking off the end. The plane spun around a fraction then the wheels hit the ground with a thud and we hurtled through the saltbush and small saplings. Then crunch. As the plane thudded to a dead stop in a rabbit warren at the base of a low dune we were thrown forward roughly in our seats.

'The bloody thing's on fire. Get out!' King and I were trapped in the back seats and could not evacuate until Barry and Peter had both exited. In the same calm manner that he informed us that his plane didn't want to fly, Peter told me that it was not smoke that I could see, but dust. 'Pig's arse!' we yelled in unison from the back seat, as we half pushed them out of the door while the blue–grey smoke intensified. We dived over the seats and slid out of the door, expecting the thing to blow up at any second.

We moved off to a safe distance while Mr Casual sauntered over to his engine bay where the plume of smoke was emanating. Flames soon engulfed the nose of the plane. Following Peter's lead we all sprinted towards the fire and started throwing handfuls of sand mixed with sticks, stones and kangaroo turds into the flames. He was not carrying a fire extinguisher.

After the fire had been quelled by the sand we looked at the charred hoses, the broken wheel guards and wingtips. The aroma of burning plastic, rubber and roo turds wafted from the grit-filled engine. King and I stood back, a bit dazed by the whole event while Barry continued to curse. He had not managed to utter a word with more than four letters since the plane had hurtled off the runway. Peter said nothing as he started trying to push the plane backwards by himself. At last Barry recovered his vocabulary. 'And what the f... do you think you're doing?' he challenged in an exasperated tone. 'We'll push the plane back to the strip and take off into the wind next time,' Peter replied nonchalantly, as if this was an everyday occurrence.

This caught us all off guard because getting back into a plane that had just crash landed, caught on fire and been showered with grit did not seem like a logical move. We figured that Peter could truck his reluctant flying lemon out later. He insisted that we had only had a rough landing and there was no reason why we shouldn't fly out.

Reluctantly we pushed the plane over the fence and back to the strip while we conjured plans to avoid flying out in the damaged plane. Although he did not use the words 'chicken' or 'wussbag', Peter made it clear what he thought of our reluctance to reboard his plane. So he hopped up into the cockpit alone and fired up the engine, unleashing a cloud of dust and shrapnel. He roared off down the runway into the wind and got airborne with about fifty metres to spare. Barry was belting out a lusty guffaw as he shook his head while King and I trained our cameras on the disappearing plane. If we were going to be stranded out here at least we were going to get a photo of the plane exploding into a fireball.

It didn't explode and after a quick circuit Peter returned and taxied up to us while we rather guiltily hid our cameras, the wingtip and other wreckage that we had souvenired. We decided to fly out one at a time to the longer strip at Muloorina Station, about sixty kilometres away. This time there was no casual looking out of the window. I stared at the speedometer as the end of the airstrip was rapidly approaching. Again we overshot the end of the runway, but this time only by a few metres. We had momentum behind us and the wind in front. This time she wanted to fly!

6
Stranded on a desert island

King George and I decided that we would reach all three of the islands in Lake Eyre South by boat, launching from the point on the shoreline that we had picked out from the air. We borrowed a small, semi-perished, inflatable dingy with a tiny outboard motor. After inflating it with a foot pump on a number of occasions we reckoned that we had found and patched all of the serious holes. The faded blue Zodiac was loaded into a 4WD and we headed out on the track to Gosse Springs.

Just as all mound springs are not the oases that they are supposed to be, King George does not fit the regal rotund image that you may have imagined. Due to his gangly build, his more zoologically accurate diving mates called him 'Weedy Whiting', another species of temperate-water fish. Although he was too tall to sit in the back seat of a Hilux he was a good mate, loved a laugh and a challenge and preferred 'King George' to 'Weedy', so that's what he got.

Neither King nor I were experienced with boating or outboard motors, so we were pleased with ourselves when after forty minutes we were halfway out to the closest island without incident. This was going to be a breeze and King was already stuck into the bottle of champers he considered appropriate for launching a craft on a voyage of discovery. I thought that rum was more a sailor's drink but vowed not to crack open my bottle until we had reached the first island.

Suddenly the choppy waves dislodged the outboard motor from its warped and cracked plywood baseboard. I found myself hanging on with one hand to the throttle of the motor as it zoomed around beating up a salty froth like a Bamix. The blades threatened to chew into the feeble rubber raft or my other hand as I lunged for the kill switch. The motor was eventually restrained and then reattached to the Zodiac and we continued our journey, just a little shaken by the experience. One and a half hours after leaving the shore I was able to have a swig of rum as we stepped onto the shore of the larger of the two islands.

With the motor off we were consumed by the noise of birds. Silver gulls, cormorants and three species of tern noisily patrolled the sky while hundreds of pelicans and swans glided through the water. Among the waders on the shore with their agitated calls was a silent ruddy turnstone, giving us one of the very few inland sightings of this coastal species. Thousands of banded stilts, mainly juveniles that had flown over 100 kilometres north from their nesting colony on Lake Torrens, sifted for brine shrimp in the shallows. Their delicate, clean feathers and improbably straight, slender beaks belied their standing as one of the waterbird icons of the harsh Australian desert. Banded stilts nest exclusively on low islands formed by the occasional filling of large salt lakes. Interestingly, had we pulled up on this same shore a million years earlier we may have witnessed up to four species of flamingos also feeding on brine shrimp. One of these extinct flamingos, *Phoenicopterus novaehollandiae*, was a miniature stilt-like bird, and some ornithologists consider the banded stilt to be a link between these miniature flamingos and other wading birds. Unfortunately the banded stilt now appears to be facing pressures that could see it join the extinct flamingos as a bizarre species that 'used' to occur at Lake Eyre.

While King dozed in the shade with the dregs of his warm champagne I hurried up to the top of the sandy mound to determine whether free water was available on the island. Although the green *Phragmites* dwarfed me I could only locate feeble seeps of brackish water. Outside the ring of tall spring-fed grasses the island was well vegetated with knee-high shrubs of old man saltbush, bluebush, dune canegrass, nitre bush and even a solitary sandhill wattle. It looked promising. Within half an hour I had done a lap of the fourteen-hectare island and turned up eight dragons, including both the Lake Eyre and painted dragons. I also saw two flocks of white-winged wrens and this offered encouragement that the island might be both large and diverse enough to support rare mammals.

I didn't have time to have a thorough look for mammal tracks or the droppings of mammals but did notice some scratchings and turds that resembled rabbit dung. A dried out rabbit leg and a warren at the base of a nitre bush confirmed my suspicion. I dropped to my knees and punched the rabbit leg into the sand, shattering the brittle white bone. How dare they make it out here to my imagined haven for native mammals! My high spirits from reaching the island and finding Lake Eyre dragons instantly departed. I was left empty. The *Caloprymnus* and all its mates were finished, the bloody rabbits had won.

I didn't have time to dwell on my disappointment for too long because the third and largest island was still a mirage on the horizon. I was keen to quickly inspect the closer island before resuming our journey. After rousing King from his siesta, I decided to walk through the thigh-deep water for approximately 500 metres to the next island while I reflected on the demise of our native fauna. I was not down-spirited for long, King George made sure of that. Although he was not the most coordinated bloke around, King was determined to start the outboard himself.

After dozens of attempts King finally got the little motor started. I was almost over to the second island, walking backwards so I could observe his antics. I'm sure it was fate and not simply bad luck that the motor was in gear. For an instant after the motor fired King straightened his gangly frame with relief. However when the boat took off he lurched forward to grab it but missed the dingy and cut his hand on the spinning flywheel. True to form when he dragged the Zodiac back towards him the propeller cut his leg and he instinctively pushed it away again as he began to hop. Now off balance the starter rope in King's left hand got caught in the prop and the engine cut-out. Basil Fawlty would have been proud. I couldn't bear laughing any more and kept walking to the second island.

This island was only half the size of the first. Although it still supported somewhat dried *Phragmites* there was no free water. Along with more Lake Eyre dragons I also found a small lizard called *Morethia adelaidensis*. It amazed me that these little skinks could colonise the island while the mainland was not even visible. There were also signs of rabbits on this island and some big white dingo turds.

Unfortunately King's theatrics had broken the cotter pin in the outboard. We were now without a motor and realised we had no chance of getting to the large western island. We also resigned ourselves to a long trip back to the shore, given that we could not even see the coast where we had left the car. I started pushing the boat through the waist-deep water but King decided he would atone for his poor form and row us back to shore. He set up the paddles and arched his wiry back. The Zodiac was propelled straight back into my face, dropping me into the salty water as if I had received a Jeff Fenech uppercut. After a mouthful of saline abuse King quickly made some alterations to his rowing style and heaved another stroke. Again the Zodiac lurched backwards towards me. Rather than turning himself around so that he faced the back and propelled the craft forwards, King had turned the little plastic paddles around the opposite way, which meant the Zodiac still came backwards, but even faster! King embarrassedly turned around, faced the back of the raft and managed to get the Zodiac heading in the correct direction. I regained my composure, caught up and jumped in.

After nearly an hour of rowing in the burning sun we looked up to see the island seemingly the same distance away as when I had jumped in. The dingy was zigzagging around because the left oar holder was broken, the front of the Zodiac kept deflating and there was a strong crosswind. It dawned on us that we were in the middle of the lake in mid-November, reasonably inebriated and drifting further from our car. We had half a hip flask of rum, a mouthful of hot Coke and bugger-all water. I couldn't believe how stupid we were. Challenging one's life expectancy was the sort of stunt performed by naïve tourists, not bushies.

Due to the curvature of the earth and the almost featureless horizon it was difficult to know where we should be heading. The only landmark we could see through the distortion of heat was Ngarlamina of Aboriginal legend, which formed the modern day Hermit Hill. We were aiming for a red patch that drifted into and out of focus, as this bare patch was the only landmark near our car when it disappeared from view on our outward leg. Coincidentally this red patch was on the escarpment that received worldwide attention eight years later following the mysterious appearance of the Marree Man landscape sculpture. Despite persistent speculation and rumours, no-one has claimed responsibility for using precise surveying techniques and earthmoving equipment to engrave this image of an Aboriginal man.

We took turns rowing, reinflating the raft and bailing out the salty water that the wind was whipping off the waves. Soon we reached a point where the water was only navel deep and we could take turns to walk and drag the stricken craft. As we got closer, the car became visible and our spirits soared.

That evening our little adventure did not make world headlines. Breaking new frontiers in outback Australia was out-billed by the dismantling of the Berlin Wall. Although we had not found a colony of mammals that were presumed extinct we had learnt a lot about the islands. Since they were small and had been colonised by both rabbits and dingoes, the possibility of them being a refuge for rare desert mammals was remote. However I had seen enough on the islands to know that it was worth returning. They were well vegetated and the larger one had permanent water. At least three species of lizards had walked, swum or rafted over from the mainland. Unique animal communities are often found on islands. Maybe there were small reptiles or mouse-sized mammals that had been isolated long enough to evolve into different species? Maybe critters that were rare on the mainland may be common on the islands? These questions rushed through my exhausted brain. I had to get back out there with traps to find out some answers.

7
Climbing up a flat lake

By March 1990 the lake had lowered considerably and after our harrowing boat experience I decided that walking to the islands was a better alternative. I easily convinced a new environmental graduate, Matthew Griffiths, that a stroll in Lake Eyre would be a good way of spending a weekend. Matthew had the enthusiasm and fitness of a Labrador puppy and hence was perfect for the task. We left Roxby before first light and by dawn had followed my car tracks from the previous trip to the shore adjacent to the islands. With the rising sun on our backs we set off on foot. The islands appeared to be deceptively close.

Matthew was towing a raft of twelve, half-metre lengths of sewer pipe with fitted caps on both ends. These pipes were lashed together with rope and were to be dug into the ground to serve as pitfall traps. Some of them contained rolls of flywire and long metal pins to construct a low fence to guide small mammals and reptiles into the pits. I was carrying four litres of water for our trip and towed a twenty litre drum of water to store on the island for subsequent trips.

The receding lake had left a mudflat that we had to skate across, dragging our loads until we reached the water's edge. Walking in the shin-deep water was even more difficult and our heavy loads had to be dragged by hand while the shallow water tripped us up. With ill-founded relief we reached the waist-deep water where we could lean over, dragging our cargo with the weight of our bodies. With the deeper water came larger waves. At first these were just a nuisance, slapping against our bodies and splashing salty water onto our faces. The waves became more of a problem the deeper we got. After being pounded for twenty minutes by half metre waves our neat little raft started looking a bit shabby and then it disintegrated altogether. The pits started racing away from us like a flotilla of little craft.

Matthew grabbed an armful of three or four pits but was then snookered as he had dropped the rope that tied these together into the murky water. I quickly unhitched the water container and tied it to Matthew who, with an armful of traps concealing his face, was feeling around with his feet for the rope. My old Dunlop Volley sandshoes, which were perfect for walking slowly and deliberately, were quickly swallowed by the glue-like mud when I charged after the pits. In the meantime, both of us were trying to keep our small backpacks with cameras and food out of the waves. I stacked an armful of traps haphazardly on the windward side of Matthew, who had relocated the rope and had begun retying the traps together. My next sortie was much farther because the four little vessels I was chasing had a fair bit of momentum. Ten minutes later we had rebuilt the raft and we looked up once more towards the island.

We resumed our walk with added vigour, determined to get into shallow water before the raft broke apart again. The walk proved to be more strenuous than we had anticipated, with the wind dragging our loads to the right and straining all the muscles on our starboard side. Two hours after the raft had collapsed we got to the shallow water near the island and were relieved to have successfully negotiated the waves. With tired legs we hauled the water container off the beach and buried it in sand to prevent the sun from perishing the container and to stop the dingoes from having a chew. We dumped the traps on high ground and after a brief rest we did a lap of the small island. Along with lizards we also saw two sets of human footprints. We had no idea who had been out here but both of us bet they didn't reach the island on foot! I counted several hundred banded stilts and red-necked avocets with their frail-looking upturned bills and racing stripes on their backs. Little button quail, which were generally an indication that there was good vegetation cover and not too many predators, were a particularly exciting discovery. This time I also saw a couple of rabbits, both pale ginger in colour as is common in the white sand around the lake.

As we returned to the mainland along our outbound tracks we could just make out a glint of sun on the car. Walking back without towing our gear we felt like baseball players who go out to hit with only one bat after warming up with three. We positively powered back, taking only three hours to return, just over half of the time it took to get out to the islands in the first place.

In addition, because the physical and mental exertion was lower, we were more aware of our surroundings. While we were lazily dragging our hands through the water we noticed that we were bumping into little fish every five to ten seconds. The density was phenomenal. It reminded me of riding a bike through a grasshopper plague or standing in a hailstorm. A year after it had filled, fish were now plaguing in the lake.

While the shore of the islands was home to wading birds such as red-capped plovers and stilts, the pelicans were the denizens of the open water. Many were gliding inches above the surface in the same direction that we were heading in the afternoon. We could see some flying to rocky islands about a kilometre offshore. Our inquisitiveness got the better of our weary legs and we slopped off to investigate.

As we approached the islands it became evident that what we had thought were rocks were in fact pelicans. The islands were low sandy rises that had been submerged a few months ago. We reached the smallest and closest of what was an archipelago of four islands. Except for a five-metre sandy fringe it was absolutely jam-packed with pelicans. The adults slowly shuffled away to reveal nests containing either two white eggs or small pink chicks. We only disturbed them for a couple of minutes while we tried to assess the scale of the colony. Much to our relief the adult birds quickly returned to their nests after we passed. There was at least one nest per square metre on the first island and the other islands appeared to be similarly packed. This was by far the biggest concentration of birds I had ever witnessed. As we trudged back to the car through the water and mud Matthew and I knew that we had stumbled onto something special.

※ ※ ※

It was not until November that Matthew and I, together with John Fewster, headed out to dig in the traps on the centre island. Since there were heaps of Johns at Roxby, Fewster went by his surname. Fewster worked as a cartographer for WMC and lived in the single persons' quarters at Roxby like myself. We shared a birthday, a similar sense of humour and a love of the outdoors, so Fewster and I did many trips together.

We were weighed down with hats and sunglasses to protect us from the glare of the now dry lakebed. We also carried three cameras, two spades, food, six litres of water and traps. Fewster and I carried between us the box of aluminium, fold-up Elliott traps used for catching mammals. The edge of the lake was muddy and slippery but we soon reached a hard salt crust. Walking was easy and I figured that this trip to the island would be a stroll. Then the underlying mud started getting sloppier and we broke through the crust every few steps, sinking shin deep into the black ooze. At first we found it amusing, particularly Fewster and I because we couldn't anticipate when we would break through. One of us would sink and this invariably placed additional weight on the other who would then break through the crust. We would then put the traps on the salt crust, lightly step out of the mud and continue our walk.

The amazing thing was that we could not predict where the mud was hard or soft or where the crust was thick or thin. There was no visible difference on the surface. John and Roma Dalhunty, who extensively studied the surface of Lake Eyre North, were also thwarted by the unpredictability of these dodgy areas that they called 'slush zones'. These zones move following floods. An area of lakebed that supports a hard crust after one flood may be a sloppy bog after another. Huge amounts of salt and silt are moved around within the lakes and indeed between them during periods of inundation. In 1974 when water flowed from Lake Eyre North through the narrow Goyder Channel into Lake Eyre South, huge quantities of salt were transferred leading to the formation of a new salt crust in the southern lake. In 1984 the tides were turned. Lake Eyre South filled to a greater degree than its northern partner and as water flowed from south to north 20,000,000 tonnes of salt was returned to Lake Eyre North.

By the time we were about a third of the way to the island we were breaking through the crust consistently. The vagaries of salt movements were a long way from our minds as we strained physically and mentally. The salt crystals were cutting our ankles and shins and black mud squirted onto our thighs. Every step was like climbing up a flight of stairs wearing heavy wet boots.

We knew that we had to keep pressing on. Matthew, the labrador, was starting to leave Fewster and me behind so we suggested that we should swap loads and that he should carry the metal box of Elliott traps. Immediately he dropped back and the three of us trudged through the endless black slop, each step sinking down to our shins. Our black footprints contrasted immaculately with the snow-white salt crust and it was only by looking back at our three tracks coalescing into a single dark line that we could see that we were progressing. We were all knocked up, the island was not getting any closer and we were mindful that we had not only to get to the island, but we also had to dig in the pits, set the Elliott traps and get back again.

We had welcomed the breeze that was keeping the flies off our faces. Except for some occasional still mornings it always seemed to be windy out on the lake. With his hands occupied by the traps Matthew was not able to keep his hat on during a particularly lusty gust and it went dancing away on the salt crust. The traps had sapped his energy and he was too stuffed to chase it. We all accepted that it was gone, which is a huge loss in the most hat-needy place on earth. Just as we were about to resume our trudge the hat stopped momentarily on a salt ledge. I decided to take off after it, only for it dislodge and roll off again. But I was committed to the chase. Black watery mud splashed up my shorts and stained the white salt sheet for about a metre around each of my heavy footsteps. After fifty metres and several more tantalising but momentary 'hat stops' I managed to grab the runaway garment. The trudge back to Matthew and Fewster took about four times as many steps as the outbound trip. My outward and return tracks had left a dramatic blip in our otherwise relatively straight tracks and served as a landmark for subsequent trips.

For the third time in a row I had underestimated the difficulty of getting to the islands. The stress was not only physical but also mental. We were not sure whether we could make it out to the island and back and were aware of the serious trouble if we couldn't. Years later I walked from Roxby to Woomera in one day for charity. The eighty kilometres was a lot longer and the physical pain was undoubtedly greater than our Lake Eyre walk. But with drinks and food every couple of hours there was little mental anguish or thoughts of perishing on the road to Woomera. Lake Eyre was very different.

Eventually we made it to the island. The sandy beach provided the perfect opportunity to rest, but it was short lived. We knew we had to dig in the pits and return to the mainland that day. We relocated the pit traps and the buried water container and topped up our bottles. As we dug the pits the mud on our legs dried leaving rattly black dags stuck to the hairs on our legs.

Setting the traps occupied us until dusk. With the last rays of the sun disappearing behind us we set off alongside the black tracks of our outward journey. The heavy high stepping quickly took its toll and even though we were no longer carrying the Elliott traps, we were sometimes only making thirty metres between rest stops. The hat blip became visible and slowly passed. Silently I remembered how stuffed we were when we had got this far on the way out to the island, and figured that we still had several hours of trudging to endure. We were not saying much now, each of us consumed by our own thoughts.

The red spot in the cliff behind the car had disappeared into the night hours earlier, and this was not a problem while we were following the dark 'train' lines that were produced on our outward journey. But when we eventually reached the hard salt where we had not left any tracks, we were faced with the task of finding a car several kilometres away on an inky black night. Fortunately Jupiter was rising in roughly the right direction so we aimed for the bright planet. We also spread out to maximise our chance of stumbling across a black-boot hole left when we had occasionally broken through the crust on the outward trip. We finally reached the car just before midnight. On a fire of old railway sleepers we quickly cooked some steaks and dried the hard rattlies on our legs. With only enough energy to chew on the steaks and not to clean ourselves, we crashed exhausted in our swags.

After a sleep broken by hairs ripping every time I moved we drove down to a bore-fed lake near Gosse Springs where we soaked the black salty dags from our legs. My reflections on yesterday's walk paled into insignificance when I remembered that it was the eleventh of November. Remembrance Day put our walk into perspective and I felt ashamed for thinking that we were doing it tough the previous day.

I headed out again to the island, this time by myself because the others had 'real' jobs back at Roxby. Following the black tracks etched into the lakebed and knowing what lay ahead somehow made the walk easier. With my mind not occupied with whether I could make it or not I started focussing on what I might find in the traps. I had resigned myself to the fact that the presumed extinct *Caloprymnus*, lesser bilby or Gould's mouse would probably not be awaiting my arrival. And they weren't. However, in thirty-five Elliott traps I had caught thirty mice. Unlike snap traps that kill mice, animals caught in

Elliott traps remain unharmed and it is always exciting to peer in to see what secrets they hold. Unfortunately, these mice were not presumed extinct, rarely seen or hitherto undescribed native mice. They were stinking feral house mice. I was shocked. How the hell did they get out here? I was surprised that even rabbits had made it! It was not only that they had managed to colonise the islands that amazed me, but their phenomenal population density. On the mainland I would normally get only one or two mice from that number of traps. And it was not only the mice that were booming on this island refuge. Some of the pits also contained up to five painted dragons each whereas a capture rate of two per pit would normally be considered high.

After some raisin bread for dinner and a scout around the island with a torch looking for geckos, I sat on a tall sand hummock to take in the ambience of one of the most remote places imaginable. I love camping out in the bush soaking up the enormity of the night sky and listening to the night sounds. In my opinion, anyone who has not thrilled to the weird whooping of the spotted nightjar or been shrieked bolt upright by a barn owl has not truly experienced the outback. Just as I was beginning to relax in my million-star resort some lights to the north invaded my solitude. They looked like car lights but as I watched them I decided that it could not possibly be a car. To the north were miles of boggy salt lake, the shore was out of sight but the lights were relatively close. Oh my goodness, all of a sudden I felt small. Very, very small.

Many people who have spent a lot of time camping in the outback have a story to tell about illuminated nocturnal aberrations. To some they are known as 'min min' lights. Bobby Hunter from nearby Stuart Creek Station tells a story about a light that rose like a large moon in the western sky. As he and the other blokes in the stock camp watched it and confirmed that the true moon had already risen in the east, a bright shaft of light emanated from their false moon. The group were terrified and dived into their swags, which are renowned for the protection they provide against astronomical phenomena and alien intruders.

I wish I had a stock camp of ringers to share my current experience, whatever it turned out to be. I scratched an arrow in the sand in the direction that the lights were coming from and hoped that my sleeping bag would provide me with the same protection as a swag. Years later, a senior Arabunna man, Reg Dodd, told me that the Lake Eyre South islands were important sites for the local Aboriginal people and that I shouldn't really have been out there. Maybe I was being punished for transgressing ancient law.

The only creatures that visited me that night were mice running around and over me. There must have been thousands on the island. The next morning I caught several more mice and lizards as I closed the traps. I filled my water bottles from the buried container and bashed my socks that had dried so hard with muddy brine that it was difficult to prise them open to get my feet in. As I trudged back to the car with the heavy box of traps I wondered about the other western island that I had not visited. It held more promise for interesting discoveries for two reasons. Firstly, it was a lot larger and larger islands invariably support a greater diversity of wildlife than small ones. This island was also closer to the mainland. Another of the 'laws' of biogeography is that islands that are closer to large landmasses typically support a wider range of species than more remote islands.

※ ※ ※

I made a couple of trips to establish how to reach the shore of Lake Eyre South adjacent to the western island. I eventually found a suitable departure point with only a couple of kilometres of menacing salt flat to cross. Gingerly stepping out onto the lakebed I was relieved to find it was hard, although from previous experience I expected my foot to break through into slop at any time. Yet the lakebed was firm the whole way and in less than an hour I was standing at the southern end of the island.

Old salt-encrusted horse turds were dotted around a flowing spring. I remembered the story told to me by my historian mate Phil Gee, that the locals used to call this place Turk Island. The story goes that a stock horse named Turk used to live on the island until one day he was mustered to the mainland and trucked down to the Kapunda horse sales. Legend has it that Turk escaped from his unfamiliar surroundings down south and made his way back to the island.

I was still poking around Turk Island as darkness fell and in the fading light a small bat fluttered past. A few minutes later I heard the loud peeping of another bat that was presumably either a white-striped mastiff-bat or a Gould's wattled bat. I was half prepared to spend the night on the island and had carried with me a loaf of raisin bread and some overalls. My mind was made up for me when the car disappeared in the dark. Although I tried to sleep a couple of times the April sand was too cold, so I wandered around occasionally jogging on the spot to keep warm. The sky was so dark and the night so quiet that the stars had a distracting presence like a television in a dark lounge room. I couldn't stop gazing and wondering how many there were. All seven sisters were clearly visible, so much so that I could even pick out the pretty ones. I played games trying to find black patches of the sky with no stars. Invariably, whenever I had found a likely patch and stared at it for long enough I could make out a pinprick of light in the dead black background. At first all the stars looked white but as the hours went past I convinced myself that some were in fact light blue or green and others, not just Mars, were yellow or orange.

Everyone knows that stars 'move' across the night sky as the earth rotates, and many times I had previously noticed constellations that had moved during the course of a night. But on this night on Turk Island with no spotlights, campfire or conversation as distractions, I actually saw them moving. I found a star that I figured was due south and checked that the Southern Cross and the handle of the saucepan kept pointing towards it as my night world spun around this tiny pinprick of light.

A dingo that appeared just before dark was not interested in such trivialities. He paced around and occasionally lapped noisily from the small pools at the spring. He trotted off and twice howled out to his mates about a crazy white man shuffling around in circles and looking at the sky. Others joined his most impressive chorus just as the eastern horizon became visible as a dark brown smudge resting on the flat blackness of the lake. Maybe my dingo mate was howling out, 'Get ready, we're about to freeze!' On cue the temperature plummeted just as the sun was rising. It was as if the sun was draining energy from the Earth in order to jump-start its own heat production. I swear that for ten minutes the sun's rays were actually cold.

On the dawn walk back to the mainland I witnessed the most amazing mirages. I could clearly see the Curdimurka railway siding and the water tower, which was strange because you can't see the lake from the top of the tower. What was even more bizarre was that the little flat-topped mesas behind Curdimurka were repeated twice upon themselves, turning them into steep pinnacles. With optical illusions like these it was no wonder that I had often been confused by distances on the lake.

The following summer I made plans to trap on Turk Island and to revisit the centre islands to search for additional reptiles. My original walking companions had left Roxby but fortunately I found a new sucker to accompany me. Darren Niejalke was another ecologist who had recently been employed at Roxby and I assured him that I had wised up. I told him it was silly to cross the deepest and muddiest part of the lake from the east and that I had no problems walking out to Turk Island. Although his main interests were frogs and reptiles Darren was now studying mound spring invertebrates, so he too was keen to check out the springs in the lake.

Although the December morning of our trip produced the hottest weather I had encountered on the lake, this western route was by far the most straightforward walk I had ever made to the islands. As we got closer we encountered a number of salt-encrusted locusts and beetles that had found the going a bit tougher than we were. On hot days the salt crust effectively 'sweats' resulting in a super-salty layer of brine forming on the surface. Any critter with more legs and less clearance than an emu or a dingo rapidly 'ices up' with salt crystals. The lakebed actually forms an important trap for locusts. Billions of salted grasshopper corpses represent trillions of eggs that don't get laid and that would otherwise exacerbate the damage of plagues.

Despite the favourable hot weather the only new reptile that we added to the list was Grey's skink—Australia's smallest lizard. While it is conceivable that some of the larger lizards made it to the islands on foot, surely this tiny lizard must have caught a driftwood raft or railway sleeper out there. The dingoes on the larger island had two small pups which we hoped would stay on the island long enough to knock off all the rabbits. But even if they did it was probably too late and the rabbits, along with the high predator numbers that they supported, had already wreaked havoc. Turk Island proved to be no more pristine than the smaller islands, although our few days of trapping confirmed that it did support more types of animals than the smaller, more isolated islands. If feral animals had contaminated even the isolated Lake Eyre South refugia, it was difficult to conceive of any unaltered havens where *Caloprymnus* or other presumed extinct mammals, could still persist.

8
Cameras can lie!

The small pelican chick agonisingly straightened its neck. Millions of prime time television viewers watched the lone bird convulse and slump to the ground; in the background a haunting sound-track played. The camera pulled back from the prostrate motionless form to reveal hundreds of other chicks, their battle for life already lost. The viewing public thought that they heard a young ecologist explain that 'All of the pelicans in Australia have gathered at Lake Eyre South.' This was a monumental incident for a bird species that attracts the affections of all but the most hardened fishermen. All of the pelicans in Australia had converged on Lake Eyre and their chicks were dying a pitiful slow death from starvation. The camera never lies.

From lounge rooms across Australia concerned viewers of *A Current Affair* were shocked into action. Somehow they found my phone number and started challenging me. 'What are you doing to help the pelicans?' 'You should be organising a food collection for the dying pelicans.' 'Why aren't you feeding them?' An ABC radio announcer rang from Sydney and told me on air that he had heard that pelicans were dropping out of the sky, and he wanted to know if they presented a risk to motorists. A long sigh on live radio may not make good listening, but Sydney commuters heard it. I then set about trying to put the story straight for the umpteenth time.

※ ※ ※

Frank Badman, 'The Botanist' at Roxby, is a massive man with a tangly grey beard flowing down untidily towards his expansive belly. Frank is a self-trained naturalist, and had been observing and documenting the natural history of the Lake Eyre district for longer than I have been breathing. His knowledge of local birds and plants far exceeded that of the 'educated idiots' who breezed into and out of 'his' country with a certificate from uni but with little local experience. Failing eyesight saw Frank switch from being an avid and disciplined birdwatcher to developing an unsurpassed understanding of the distribution and status of plants throughout northern South Australia. Although he could be a difficult bloke to work with, Frank was invaluable while I was becoming familiar with the local environment.

Big Frank had not looked too interested when I rushed in and excitedly told him about the enormous pelican colony that Matthew and I had stumbled upon at Lake Eyre South. Typically Frank would have replied with 'Once I saw 800,000 of them at Whatchymacallit Bore' or 'Five million of them nested at Lake So-and-so in 1974'. Because he didn't trump my observation Frank had confirmed to me that this pelican colony was indeed special. I also questioned Dave Paton as to who I should notify. As my Honours supervisor I knew that Dave was only slightly easier to impress than Frank, but he agreed that the colony was 'significant' and suggested that I speak with Max Waterman.

Max was a serial bird bander who had fostered Dave's interest in banding birds when he was a kid, so I felt as if I was Max's ornithological grandson. Bird banders identify individual birds with numbered metal bands that are fixed around their legs. For thirty years Max had been one of the most prolific bird banders in Australia. As well as coordinating banding in South Australia he was also directly responsible for banding nearly 400,000 birds. This is a record of Don Bradman proportions! Max specialised in colonially nesting bird species that he could band thousands at a time. He had nearly cracked the 100,000 mark for crested terns and had banded in excess of 60,000 cormorants. Clearly the Lake Eyre South pelican colony was right up his alley. Max was immediately itching to come up and band as soon as he could round up some fellow pelicanophiles.

I was keen to meet Max and to study the pelicans that Matthew and I had found. Working with a nesting pelican colony of this size was a unique opportunity that I did not want to miss. Maybe we could collect some clues as to how they located the lake full of fish *en masse* nearly a year after the rains. My only problem was that I had my old Belmont station wagon packed for a pilgrimage to Bow River in the Kimberleys with a mate, James Nagel. As we drove off with Cold Chisel declaring that 'The money I save won't buy my youth again', I resolved my dilemma by reconciling that Matthew could show Max how to get out to the lake if Max got his act together before I returned. Pelicans nest for about three months, so there would still be plenty of activity in the colony in a month's time.

Max had got his act together. After changing my sixteenth flat tyre for the year when I returned from the Kimberleys in mid-April, I raced to Lake Eyre to immerse my life in pelicans for the next month. Speeding past the Gosse Springs I found that where there had been a faint tyre track from our two previous visits, there was now a well-defined track. A tent city had sprung up on the shore adjacent to the pelican colony.

A motley crowd appeared from their crouched positions out of the wind behind low samphire bushes. Their large jovial spokesperson greeted me like a favourite nephew and announced that he was Max Waterman. Max easily filled out his clothes that had clearly come out of his 'field work' drawer. He didn't strike me as the sort of bloke who could single-handedly band hundreds of thousands of birds and the thought of him walking out to the islands, let alone scampering around after a flock of pelican chicks intrigued me. I was introduced to Max's team of twelve who ranged in age from about seven to seventy years. One of the group was the ornithologist Kim Lowe.

Together with Max, Kim estimated that although the numbers had dropped since we discovered the colony five weeks earlier, the old nest scrapes and the 30,000 remaining chicks suggested that the original colony size was around 100,000 pelicans. Kim estimated that this represented eighty to ninety percent of all of the pelicans in Australia! The enormity of our discovery became immediately apparent.

Whenever Kim stopped for a breath Max would fill the void by repeating *ad nauseam* how fantastic it was that he had the opportunity of banding this colony. According to Max I was a great bloke and WMC was a model corporation. The pelican islands were the most spectacular sight he had ever seen. No-one else said anything at all.

Max told me that today was a rest day, which I thought was a term reserved for the day after the third day of a test match. I had certainly never heard of a rest day on a biological field trip! I reasoned that it was a day that Max used to savour the thrill of having so many legs to band and so many attentive ears to tell of his banding exploits. While I listened and questioned Max about his plans everyone else politely melted away, perhaps relieved that they were not the current recipients of the relentless Waterman enthusiasm.

Despite his bumbling yet bombastic nature, I liked Max. Later I learned that he had received an Order of Australia, the only one awarded for bird banding. But Max did not seem to band birds for scientific acknowledgment or status. To my knowledge he had never written an article on his work. Because Max was a smooth-talking salesman in his other life I could understand his obsession if he was making money from his bird banding, but the opposite was true. Max didn't need a scientific or economic rationale for devoting most of his spare time to bird banding. He had started banding soon after the national banding had commenced in the mid-1950s. Prior to that he had been an avid egg collector until Herb Condon, the curator of birds at the South Australian Museum, had steered him towards the less destructive and more informative art of 'collecting' that he subsequently embraced with passion.

Max was an interesting phenomenon but I was here principally to learn and try to understand the pelicans. The banding could tell us where they departed to and how long they lived. But other questions remained unanswered in the banding camp on rest day. What were the pelicans eating? How did they locate the islands? Were dingoes, eagles or gulls preying on the chicks? To get a feel for these issues I headed out towards the colony that was approximately 500 metres offshore. Due to the lakebed's appetite for footwear I walked into the water barefoot. The stinging cold of the water was soon forgotten by the cutting of a thousand needles of salt crystals that had formed on the lakebed.

While I was wading out to the islands, I was treated to a most spectacular and unforgettable sight. A line of evenly spaced dots, like fenceposts, was gradually moving across the horizon. Unlike fenceposts this line of dots was slowly undulating, like a giant caterpillar inching its way along a notional line where the horizon should be. They must have been getting closer because after a couple of minutes I could establish that the dots were actually pelicans flying in formation just a few centimetres above the lake surface. The pelican in the lead would flap its wings a few times, and in doing so would lift a metre or two before gliding back down and skimming the surface for perhaps 100 metres or more. In turn the formation behind the leader would follow so a wave would gradually pass back down the line. As I looked around I could see several of these flocks of impeccably spaced pelicans returning to the colony. The longer lines, those with more than forty or fifty birds, would have a couple of waves moving through the formation at any one time. The precision and control demonstrated by these huge birds was astounding.

The lowering water level had caused two of the 'original' islands to coalesce and crèches of pelican chicks crowded the newly emerged spit. The island that Matthew and I had first walked to was now bare. This island, like others in the group, essentially looked like a low white sandbar. However, vague concentric rings in the underlying limestone indicated that it was an extinct mound spring, perhaps occurring on the same fracture that provides a conduit for water from the Great Artesian Basin to form the more impressive islands in the centre of the lake. A month ago this island was a stinking squawking crush of white fluffy chicks. Now only seven chick carcasses were found on the whole island.

In smaller colonies, where materials are more accessible, pelicans decorate their nests elaborately and sometimes even construct platforms. No such luxuries were available on these islands and the most ornate scrapes in the coarse white sand were lined with a few dried pelican turds, and maybe an occasional feather or stick. The scrapes were typically less than one metre from their neighbours and by estimating the size of the island it was apparent that it had supported approximately 15,000 nests. Most pelicans lay two eggs and we observed that in the early broods when food was abundant that the two chicks generally survived, at least until the point where they shuffled away from their nests to form crèches. This island had clearly seen an exceptionally successful proliferation of pelicans.

The situation on the other islands was not so clear-cut. Through my binoculars I could make out dense pockets of adults apparently sitting on nests, and crèches of nearly fledged chicks. A continual shuffling, grunting and guttural bellowing resonated throughout the colony as chicks mobbed adult birds that were returning from their feeding sorties. Smaller chicks that were unable to move craned their ungainly heads towards returning adults. Young chicks are fed up to eight times a day so there was always plenty of activity in the nurseries. The smallest scalded-looking chicks with swollen bellies and skin two sizes too large looked like creations that had been discarded from the Muppets due to their gangliness and improbability.

The morning after I arrived at the tent city on the shore of Lake Eyre South I embarked upon my first pelican muster under the command of Major Max. Not everyone could fit in the little dinghy so I walked. The salt water stung the cuts on my feet from yesterday and every additional step added more. The deeper water provided some respite because the salt had not crystallised and the lakebed was still muddy. Once we reached the designated location on the island, long fishing nets were suspended from star droppers to form a makeshift yard and corral. While we were building the yard, most of the adult birds and newly fledged chicks flew from the island. The unfledged chicks had conveniently grouped themselves into crèches of similar-sized birds, the largest of which shuffled towards the edge of the island and prepared to take to the water. On Max's command two parties of banders flanked the opposite shores of the island trying to keep as many of the large chicks on the island as possible. The faster 'out-riders' from each of the banding parties joined around the back of the pelican crèche that we were mustering, while the slower walkers moved in along the flanks, waddling the birds towards the pen. We were careful to target the crèches of larger birds and did not drive them through nests with eggs or newly hatched chicks. Young or weak birds were drafted off the back of the mob to prevent them from being squashed or exhausted.

Once the pelicans had been mustered into the corral, the banding team split into two groups. 'Catchers' would enter the pen and grab two birds at a time by their upper wings and take them to the perimeter of the yard. The 'banders' then closed the individually numbered bands by hand. Within three musters over a couple of hours we banded about 500 birds, one eighth of the juveniles that were banded from the colony.

The size of pelicans can be easily underestimated until you are crammed in among defiant large birds, particularly when you bend down to grab their wings. From head to tail they measure over one and a half metres and then you have to add that enormous bill. Adult birds weigh up to six kilograms. One chick that departed from Lake Eyre landed outside the Andamooka Police Station, 120 kilometres further south, a month or so after we banded him. This young pelican who was nicknamed 'Percy' was reluctant to leave the Andamooka cop shop so they brought him around to the Olympic Dam Environmental Laboratory, where I worked. Several locals donated fish that barely interrupted Percy's insatiable appetite, for a pelican can eat approximately two kilograms of food a day! When Percy arrived we were also looking after a wedge-tailed eagle that was convalescing after a collision with a car. To the amazement of Pat Garland, our jovial Welsh chemist, Percy waddled straight up to the wedgey and bullied it away from a rabbit carcass. In a flash the whole rabbit—fur, claws and all—disappeared down Percy's throat while the world's fourth largest eagle looked on powerlessly.

Percy was at full adult size and dwarfed the wedgey that was just over half the weight of an average pelican. The pelicans' long necks and half-metre long beaks enable them to attach themselves onto almost any part of your body. On several occasions while we were banding them I found my head inside the huge balloon of a beak. Along with the other catchers I scored long red scratches on my legs, arms and face which we were reminded of when we hit the salt water on our way back to camp.

Another characteristic of pelicans that is lost on the casual observer is their smell. One hundred thousand fish-eating adult birds could hardly be expected to smell like a bed of roses, but the source of most of the smell was the chicks. When stressed by being herded into a pen where they are handled and inverted, the chicks disgorge their foul-smelling stomach contents. Likewise, a young pelican who is struggling to find a few extra yards in order to make it to the 'safety' of the lake, or one who might be trying to make its first excursion into the air will readily jettison its payload. The result of this spew-fest, when combined with the pâté produced by the pelicans' other end, was an almost overpowering odour reminiscent of fish sauce that has exceeded its use-by date.

All this regurgitated fish not only exercised our noses but also enabled us to determine where the pelicans were feeding. For the past year I had been monitoring changes in the salinity and critters of Lake

Eyre South as it dried. At no stage had I sampled the catfish, yellowbelly, black bream and yabbies that we picked up from the rookery. In fact, by the time we were banding the chicks, the Lake Eyre hardyheads and bony bream, which both the pelicans and I had encountered in their millions months earlier, had disappeared from the lake samples. Several rings of these dead sardine-like fish around the shore revealed their fate.

Sudden changes in water temperature or salinity, caused by the wind mixing the surface and bottom water, may have been responsible for some of the early fish kills. Alternatively, algal blooms that occur frequently in the lake may have depleted oxygen levels causing the fish to literally suffocate. However, there is little doubt that the final crunch for the fish came when salt levels increased to approximately three times that of sea water. The rings of dead fish that I was observing were not unexpected. When Lake Eyre dried in 1974–75 an estimated 20,000,000 hardyheads and an equivalent number of bony bream were washed up in four separate rings of dead fish on the shore.

The massive increase and demise of the hardyhead populations that had supported the initial stages of the pelican colony seemed to be a waste. Why should they breed like crazy only to feed pelicans or litter the shore when the lake became too hostile? Surely once they have made it into the lake their genes have reached a dead end. Contrary to popular belief, hardyheads and bony bream do not live in the mud under the salt crust, nor do they lay resistant eggs that hatch during subsequent fillings. Rather, they wash or swim downstream into the lake from mound springs or permanent waterholes. Although moving into the lake appeared to be suicidal for these little fish I later discovered that the news was not all bad for fish in the lake.

If rain falls when the lake is full of fish a mad scramble ensues whereby the fish congregate around the sources of the inflowing fresh water. In their hundreds of thousands they swim upstream like miniature salmon. In the lake they have had the opportunity to feed on the abundant plankton, breed and mix genes with populations from other areas within the catchment. Now their ambition is to make it back to a permanent spring or waterhole. By sheer weight of numbers enough will make it to safety to keep the species alive, the others succumbing to road culverts, herons or desiccation if they do not make it to a waterhole or spring.

The fish and yabbies that the pelican chicks were regurgitating in April had come from the Cooper and Warburton river systems, several hundred kilometres to the north. From our camp we watched adult birds in formations of eight to nearly 1,000 peel away from the rookery during the morning, and disappear from sight to the north. When the parents returned after a day or more from a successful fishing foray they somehow located their brood. What followed was a bizarre sight no matter how many times it was witnessed. The fish that were often big enough to throw into a pan were regurgitated whole for the hungry chicks. Because they lacked fishing knives, teeth or even claws, the chicks swallowed their food whole. Snakes are proficient at swallowing prey larger than the width of their throat, pelicans are not. Perhaps the fish blocks their air supply causing oxygen deprivation and forcing the chicks to violently convulse and bite at their wings and legs before collapsing motionless on the ground. After a while they regain composure and stand up as if nothing had happened. It was a moment like this that was filmed and appeared on *A Current Affair*.

After pushing a reporter and her entourage around in boats because the water was too salty for their delicate tootsies, and after providing them with endless information on the success of the colony, I was offended that they had turned a magnificent story into a cheap, sensational lie. For their program they had edited my voice to suggest that all of the pelicans in Australia were there and with their footage they implied that most of the chicks had died. In contrast we reckoned that the vast majority of chicks had successfully fledged. I was incensed that they would cut and dub my 'voice-over' to satisfy their bullshit story-line at the expense of the truth.

I rang the TV station the following morning and told them that I was disappointed at the negative slant that they had put on the story and that I felt my professional credibility had been blemished. They were surprised at my outrage. I guess they had not anticipated any complaints arising from a story about dying pelicans. They were wrong. Then, to compound my anger, I was told a few hours later that they could not comment on my 'allegations' because they were unable to find the footage from the previous night. Finally, a week later, the producer tried to palm me off. He said I was overreacting to a trivial incident. Geoff Whitham, WMC's wise old lawyer, told me in legal terms that the media could be unscrupulous bastards and unless I had an independent video of the interview I didn't have a chance of getting the retraction I was asking for. I promptly rang Channel Nine and told them that I had legal advice that they had blemished my credibility and that they were liable to rerun a correction and an apology.

The producer chastised me for wanting to interfere with the story-editing, that was his job. The important story I was informed was that there were a hell of a lot of pelicans there and it didn't matter whether it was '*all of the pelicans in Australia*', or whether 'Dr Kim Lowe has estimated that 80% of *all of the pelicans in Australia*' had congregated at Lake Eyre South. I was told that the viewing public wasn't interested in exact numbers and their show had conveyed the huge size and significance of the colony. It apparently did not matter if the colony was an unprecedented success or a pathetic failure.

Finally, after extended haranguing, they ran an update on the pelican story where the same footage was used but a less contrived voiceover in some way corrected the mistakes. Since that event I have felt sorry for any person who has had a foot and camera shoved in their door by an unscrupulous reporter under instruction from a manipulative producer. Ever since the shonky pelican story I have refused to watch or listen to anything that this particular journalist was involved with, even when she moved to a prime time slot in Adelaide. As well as putting me off 'sensationalist' journalism, the distorted and untruthful pelican story galvanised a misplaced public sentiment for the pelicans that were enjoying a phenomenally successful breeding episode.

※ ※ ※

In late April 1990, over a month after the successful departure of most chicks, nearly 4,000 eggs were laid on the barren island that had hosted the first pelican nests at least four months earlier. This caused great excitement in the pelican camp. Max recognised the renewed nesting activity as a signal that the pelicans knew it was going to rain again. The lake would freshen up, the fish would return and the phenomenal colony would continue. The media also got excited. Surely if pelicans were able to read climatic patterns well enough to congregate at Lake Eyre nearly one year after rain, they could also predict future rainfall patterns.

It did not rain and the lake became shallower and saltier. Pelicans usually incubate their eggs for just over one month, so by mid-July we knew that the eggs that had not hatched weren't going to. When we inspected the colony we found only eleven chicks and it appeared that the flocks of crows that we had been watching from the mainland had accounted for many of the unhatched eggs. The incredibly salt-tolerant samphire bushes were now reclaiming the islands. We didn't hold out much hope for the few remaining chicks. It takes three months for chicks to fledge and it was unlikely with the colony now nearly completely disbanded, that the few remaining parents would continue their flights to the Cooper or Warburton rivers for long.

In August I returned to the islands to tally up the dead chicks and to assess the success of the colony. In the wash-up it appeared that in excess of 88,000 chicks successfully fledged from the estimated 104,000 eggs laid. In the space of a few months the population of Australian pelicans had almost doubled. Admittedly, some of the chicks that left the island probably perished *en route* to other feeding grounds. Some starving chicks banded from our colony were found as far north as 1,353 kilometres away in Normanton in Queensland, and 1,200 kilometres north of Lake Eyre South at Brunchilly Station in the Northern Territory. In spite of the post-dispersal mortalities 'our' pelican breeding event was the most successful ever documented.

9
Giant white bees

Most of the pelicans in Australia had converged on Lake Eyre South over a period of several months. There we found a pelican that had been banded south of Perth and birdwatchers from around the nation had indicated that 'their' pelicans had departed their typical haunts during this time. How did these pelicans know that there was a smorgasbord of food at the lake in early 1990? Furthermore, how could the pelicans predict the window of opportunity offered by the barren islands that were submerged when the lake filled and that were linked to the mainland soon after they departed? The failure of the last 2,000 nesting pairs concerned me because the illusion of the environmentally 'tuned-in' pelicans was shattered. It became apparent that pelicans can't predict the rain weeks or months into the future. I also suspect that they are unable to predict when the fish stocks are high and when the islands are suitable for nesting, despite what many people confidently told me.

Some ornithologists suspect that the pelicans can detect sonic or even ultrasonic vibrations of massive storm fronts from thousands of kilometres away. This may or may not be the case but a storm does not mean that the lake will fill. A huge storm in Queensland may not put any water in Lake Eyre for many months. Even if the lake fills from local rains, fish numbers will not immediately be sufficient to sustain huge breeding colonies of pelicans. Australian pelicans need to eat approximately 500 Lake Eyre hardyheads, or slightly less bony bream, per day—a formidable task when the fish are scarce. When feeding large chicks their food requirements double. Therefore the birds must wait in their normal feeding areas for several months after the big rains while the fish in the lake breed up. Do they have some way of calculating dates so they can plan their arrival to coincide with peak fish populations? I strongly doubt it.

Maybe the pelicans can smell the fish in the lake? Several bird species including starlings, swifts, pigeons and petrels use their highly developed sense of smell to locate their homes from many miles away. Researchers know this because, believe it or not, they have cut the olfactory nerves of some individuals and demonstrated that they are less proficient at finding their way home than their counterparts with 'noses' intact. Despite the proven ability of birds to navigate by smell over smaller distances, I remain sceptical that pelicans can smell the burgeoning fish supplies in Lake Eyre from thousands of kilometres away, especially from upwind. The phenomenal long-distance movements of migratory birds are triggered not by air vibrations or smell but by seasonal cues such as changes in temperature or day length. But the availability of fish at Lake Eyre is not seasonal and the timing of peak fish numbers is probably dependent upon the time elapsed since rain. Once again, animals living in the outback must develop mechanisms for coping with, and taking advantage of the erratic rains.

There must be a way! One hundred thousand pelicans could not just appear in the centre of Australia by chance. For a feasible explanation we can turn to one of the first mechanisms discovered to understand the colonisation of food resources by animals. In the 1940s an Austrian named Karl von Frisch received a Nobel Prize for discovering how bees will share information with the rest of their hive. School kids now learn that a bee that successfully finds a particularly rich flower patch performs a waggle–dance that conveys to the others the direction and distance to this area. By doing so the bee colony saves time looking for food and thus maximises its health.

I reckon that a pelican colony may be the equivalent of a beehive and that in Australia Lake Eyre is occasionally the rich flower patch. Scouts from different pelican colonies throughout Australia independently locate the fish and nesting islands, and then encourage their mates to travel for thousands of kilometres to take advantage of them. What we witnessed at Lake Eyre was not one 'hive' but an agglomeration of several distinct 'hives'. I contend that pelican colonies from around the continent independently followed their scouts to Lake Eyre South over a couple of months.

This is a bit far-fetched, I hear you thinking. Has John followed these scout pelicans to Lake Eyre and back? No, I have not. Has he observed them doing a fishy waggle–dance? Again I have not; to my knowledge no-one has. But hang in with me and I will explain. In order to extrapolate the waggle–dance theory to a pelican colony I first have to convince you of five truths. If you can accept each of them as being a reasonable interpretation of observed facts I reckon that I might have you with the whole theory.

Let's start with the first question: do pelicans send out scouts? Between March 1989 and February 1997 I recorded pelicans in the Lake Eyre South region in over twenty percent of the months, even though food was only readily available in Lake Eyre for a few months prior to April 1990. Many of these records were of small flocks flying over when the lake was dry. Furthermore I frequently see small groups of pelicans circling or gliding over the desert miles from any lakes. Occasionally a few turn up on a pastoral dam. Sometimes pelicans are even seen standing at the edge of a road or their big webbed footprints are found waddling over a dry sandhill.

Some of these records may be attributable to dispersing individuals, particularly juveniles, from nesting colonies. These individuals may be lost and desperate. However, the relatively regular records of pelicans apparently in good condition suggest that adult birds may frequently visit the region. I interpret many of these visits to be scouts checking out the local conditions. When food gets tight or parasites build up at established colonies a few adventurous individuals may decide to check out conditions elsewhere.

Sending out scouts is one thing, but the second big question is whether the scouts can operate over a continent as immense as Australia. Max Waterman's banding suggests that they can. Five of 'his' birds were recovered over 2,500 kilometres from their banding location and a further twenty-six were relocated at least 1,000 kilo-metres away. The secret to their long-distance movements is their huge 2.5-metre wingspan. I have watched pelicans labour into the air then powerfully fly upwards until they reach a thermal. From there most of their work has been done and they let the rising air take over. Flocks effortlessly spiral upwards in these thermals scarcely flapping their wings until they disappear into the blue wash of a cloudless summer sky. These spiralling 'kettles' of birds achieve heights of up to three kilometres above sea level where they can be seen from the windows of domestic airliners. The pelicans then pull out of the thermals and glide, probably for hundreds of kilometres, without raising a sweat.

A physiologist might be able to tell us the equivalent energy expenditure of a bee frantically flapping its four wings back to its hive and a pelican gliding around on thermals. I expect a kilometre for a bee would equate to 1,000 kilometres for a pelican. The other advantage that a pelican holds over a bee is its large size. Small birds and insects have to feed frequently to maintain energy levels, but a pelican's large size permits it to travel for several days without feeding. If tiny migratory birds can travel thousands of kilometres without feeding or hitching rides on thermals, a trans-Australian flight for a pelican would be a cinch. Within a few days it is conceivable that a pelican could cruise from the coast to Lake Eyre and back. The job is even more of a stroll if a small flock moves together, sharing the hard work at the head of the line or, when they are really serious, at the apex of a V-formation.

Ok, maybe pelicans are able to scout to and from central Australia. But the third question is why a pelican that has flown halfway across

The Old Ghan railway at Curdimurka was damaged by the record floods of March 1989.

Tracks left after trudging through the saltcrust in search of rare wildlife on the islands in the centre of Lake Eyre South.

Dead pelican chicks make more controversial media stories than tens of thousands of healthy ones. (Photo: T Lewis)

The author on Turk Island about to walk to the barely visible islands in the centre of Lake Eyre South. (Photo: J Fewster)

Some of the pelican chicks banded at Lake Eyre were subsequently found over 2000 km away in northern Queensland.

Over 80,000 chicks were fledged in the largest recorded Australian pelican colony at Lake Eyre South in 1990.

The flooded ring of tea-trees in March 1989 indicates the previous high water mark at Coorlay Lagoon, Roxby Downs.

In the early 1990s, swans and coots bred in their thousands on lakes near Roxby, such as Lake Koolymilka.

Two size classes of young white cypress pine trees are rare examples of successful recruitment in the Roxby region over the past century.

The world's most venomous snake, the inland or western taipan, led to the discovery of an amazing number of rare mammals inhabiting the Moon Plain, near Coober Pedy.

Gibber dragons, like inland taipans, are recently discovered reptiles that live in the most barren stony plains of the outback.

This spunky Pernatty knobtail, with its characteristic skinny tail, is from the isolated dunes south of Woomera.

The boldly marked Arcoona dragon may be discouraged from inhabiting the rocky cliffs around deep lakes by tern predation.

The remains of small dragon lizard carcasses at a gull-billed tern colony at Coorlay Lagoon.

The fat tail of the smooth knobtail may enable it to inhabit regions that are too dry or too cold for slender-tailed knobtails.

Predation of rare outback wildlife by gull-billed tern and silver gulls may increase with expected floods caused by global warming.

the continent and found a fishy banquet, would want to fly all the way back and tell its mates? Of all of the five questions, this is arguably the most difficult to understand, but fortunately von Frisch's bees can help out again. Bees get their mates in on the action because they are all related and the success of the hive is more genetically important than the persistence of a gluttonous, sexless worker. But pelicans are different. They all contribute their own genetic stock and hence are not bound by such altruism as colonial insects. Or are they?

Attracting colony members back to a remote feast may simply be an ingrained trait that helps related individuals, like in a beehive. But there are a number of other rationales that could also explain their gregarious tendencies. By keeping the colony together a scout will maximise its chance of reaching another favourable area once the fish supply in the lake has been exhausted. It is also much easier for pelicans to locate those thermals which help them to move long distances when they can spot a 'kettle' of mates swirling skywards, rather than wasting time and energy finding the thermals themselves. Fishing can also be more successful for pelicans in a flock rather than alone. Pelicans sometimes cooperatively muster a school of fish within a closing circle of birds. By attracting a large colony of pelicans to its food source a successful scout would also increase its choice of mates. Lastly, a chick being raised in a colony is afforded more protection from predators and the elements in a crèche of other chicks, rather than remaining alone and exposed while its parents are off feeding. Therefore, at least sometimes, pelicans can benefit from forming and maintaining colonies.

Bernd Heinrich, an American bird-behaviour scientist and fanatical raven watcher, has found that raven roosts are 'mobile information centres'. The ever-changing occupants of these roosts can learn the location of a remote food supply simply by following their successful roost mates. Similarly, if we catch up with a mate at the movies it is easy to work out if they have just eaten garlic bread and a pizza with the lot. By the same reasoning a pelican is sure to arouse considerable interest if it returns to its colony with chronic fish breath and a girth that indicates that it has been in a 'good paddock'.

Pelicans are social birds that use a complex set of postures, movements and sound signals to communicate with members of their colony. In the same way that European swans signal the quality of their feeding grounds before leaving their roosts, Australian pelicans may also communicate the presence of rich remote food supplies to members of their colony. This 'communication' may not be as complicated and rehearsed as a waggle–dance.

You may now be thinking of the fourth question: why would the colony decide to move *en masse* from their regular, reliable food source into the outback? Such a relocation from established rookeries could enable their depleted fish stocks to recover, may allow the pelicans to move from an area where parasite loads have increased, and may enable them to interbreed with other colonies. Even though a coastal location may provide more permanent food resources, coastal nesting islands may be limited or in danger of destruction by dogs, livestock or humans. These areas may also support high gull populations that scavenge for food regurgitated for the chicks that are too young to feed directly from their parents' beaks.

The fifth and last line of evidence that I will provide to support the scout hypothesis is that the Lake Eyre South pelican colony, in keeping with subsequent colonies observed on Lake Eyre North, was made up of distinct sub-colonies. If all the pelicans had arrived at Lake Eyre from around the country at the same time, I would suspect that they were all responding to the same visual or olfactory cue. However this clearly was not the case. At least four distinct nesting attempts were initiated on the 'pelican' islands in early 1990. Freshly laid eggs adjacent to patches of newly hatched chicks, and other regions of month-old chicks suggests that large groups of pelicans arrived at different dates. Therefore several localised stimuli rather than a single broadscale environmental cue probably initiated the colonisation of the islands.

I suggest that scouts periodically leave established feeding and breeding grounds and engage in reconnaissance flights like giant white bees. Their 'reccies' may be more frequent after inland rains or when conditions at their colonies become harsh. We don't know. We also don't know whether they follow a predetermined route or simply glide around as the winds take them. In the event that they locate food supplies that are sufficient to support breeding, these scouts then return to their colony and lead them back. So many pelicans flew to Lake Eyre South in 1989–90 because this was the best place in Australia for breeding at the time. In years when rain and fish stocks are more dispersed, such a high percentage of the nation's birds would be less likely to descend upon one small isolated archipelago.

Almost a decade later four small pelican colonies, each numbering less than 1,000 pelicans, nested on islands in Lake Eyre North, while another group of 400 birds nested on the islands in Lake Eyre South. There had only been water in the lake for about a month, so the fish had not bred up and the water levels were already dropping quickly. The fact that these small groups of pelicans were not joined by multitudes of others indicated that suitable food and nesting resources were available elsewhere, or that only a few scouts reckoned there were enough fish in the lake to warrant enticing their colony. The timing and numbers of birds matched the fish numbers in the lake. I can't wait for the next filling, to test whether the numbers of pelicans and their breeding times also conform with the unique timing of peak-fish numbers.

Maybe the whole story of the ecology and movements of Australian pelicans will only be revealed by radio-tracking dozens of these birds from many colonies over several decades. Or maybe we will need to avail ourselves of the newfangled genetic techniques that are being used to unravel the movements of migratory waders. Until such detailed research occurs we will not have the confidence to teach our children about the waggle–dance equivalent of our pelican populations.

10

Ducks in the desert

'It's easy, you just stand up,' shrieked the young voice from behind the boat. Debbie Oag was a natural, as if she had been born on water skis instead of growing up near a dry lake in the desert. 'Watch this,' she shouted as she effortlessly glided from side to side with Dad smiling proudly and big sister Lorena itching for her turn. When my turn came I learned that Debbie had lied. A 75-horsepower facial enema confirmed that water-skiing wasn't as easy as it appeared.

David Oag had bought the boat just after the 1989 rains, when the floodwaters inundated the old Arcoona homestead near Woomera for the first time since it was built in the 1870s. A second-generation manager on the Kidman-owned property, David knew how special it was to have a freshwater lake lapping at the doors of his new homestead and was determined to enjoy it with his young family. It was not only the long-time pastoral families who were taking advantage of all the water in 1989. The lakes became the playgrounds for everyone in the district. All of a sudden more than twenty speedboats appeared at Roxby. Paragliders whipped into the air above the lakes while scuba divers collected yabbies under the surface.

A series of ten major lakes, each at least one kilometre wide, and many smaller swamps were filled by the run-off from the Arcoona tableland between Woomera and Roxby. Due to their common origin from this raised gibber plain I called them the Arcoona lakes. The lakes were all different depths and salinities. Shell Lagoon was a shallow salty lake that was twice as saline as seawater before it dried in September 1991. By contrast, Arcoona Lake lasted until 1995 and Lake Campbell, on Roxby Downs Station, was still fresh when it dried four years after the rain. Unlike the other lakes that gradually cleared as the salinity rose, Lake Campbell stayed muddy until it dried.

Once dry, the sediment that washed in during the initial flooding was whipped up by strong spring winds. The finer dust probably ended up as topdressing for farms in the eastern states but the coarser, heavier material dropped closer to the lakes. Dust storms formed the sand dunes that hug the shores of the larger lakes. As the sand gets older, or further away from the lakes, rusting of the iron compounds changes it from white to yellow to the deep orange that characterises the old sand dunes in the region.

Fish were introduced into some of the lakes but they proved too elusive for anglers. However, pioneering local mechanic, Bluey Lavrick, pulled out some huge six-kilogram yellowbelly and catfish using his 'square hooks' nets within eighteen months of introducing them. The Murray cod and trout that were introduced presumably fell victim to the waterbirds or other fish because they were not recaptured. Apparently fish did so well when placed into Lake Koolymilka in the mid-1970s that pelicans nested there in 1976. It seems unlikely that this colony of pelicans found this unexpected food bonanza by chance! I reckon that scouts were probably involved again. Unfortunately in 1989 Lake Richardson filled with feral carp that overflowed from the flooded Woomera sewer ponds.

Perhaps the most novel use of the lakes was by Zolly Farenci, who irrigated a large patch of watermelons adjacent to Coorlay Lagoon. Zolly, a local driller, was no doubt inspired by the success of paddymelons that grew in large quantities around the lakes. Paddymelons are related to watermelons but unfortunately their round, softball-sized fruit are absolutely inedible. With the exception of the inappropriately named native apricot, paddymelons have the most acrid, face-twisting taste imaginable. Zolly's watermelons, carefully fenced from rabbits and sheep and irrigated from the lake, thrived initially but were beaten by the blistering north winds of summer.

Like nearly everyone else at Roxby I loved visiting the lakes for a dip or a ski but I was also interested in the bird life that they attracted. Banded stilts, gulls, terns and pelicans had already swarmed to nearby salt lakes Torrens and Eyre, like bargain hunters to post-Christmas sales. However these smaller, fresher lakes were initially almost devoid of water birds. Then gradually, as the salt lakes and shallow swamps in the surrounding country dried, birds congregated on the Arcoona lakes. Increasing aquatic plant and invertebrate numbers provided a food source for the burgeoning bird population. I could not help but wonder what secrets these largely unstudied lakes might hold. I had found myself a mission.

Before these lakes filled I enjoyed admiring water birds for what they were. Unfortunately though, the desire to understand these creatures obscured the sheer enjoyment of watching them behind nagging questions. Why did musk ducks accumulate on Coorlay Lagoon but not on any other lakes? Will they breed there? Why did coots hang around with swans? The questions were as endless as they were distracting. Scrutinising birds' behaviour is not as relaxing or enjoyable as watching ducks gliding through their reflections on a mirror lake. Ignorance is indeed bliss with birdwatching. It was ironic that I was learning how to identify over fifty species of waterbirds in one of the driest parts of the continent.

Ducks proved easy to sort out. Before long I could identify them all by the way they flew or sat in the water, which is also a prerequisite for anyone gaining a duck hunting license. Even when ducks can be seen in more detail their names often bear little resemblance to their appearance. Both grey teal and black ducks are brown in colour. The tiny pink patch on the head of the pink-eared duck is one of the least obvious features of this boldly striped, long-billed duck. Flying style is also diagnostic. Hardhead flocks usually flew higher and for longer than other ducks when disturbed, the white down the length of their wing distinctive. The bizarre looking musk duck, which appears too heavy to float, and the blue-billed duck rarely fly at all but dive under water when alarmed. Although their freckles were only apparent from close range, freckled ducks were readily identified flying with their heads slumped down heavily as if their necks were broken.

These primitive freckled ducks, which are actually closely related to geese, are restricted to Australia and are one of our most threatened bird species. Freckled ducks always arouse interest, especially when the males have red bills, which is supposed to indicate that they are in breeding condition. I observed hundreds of red-billed males over several years but unfortunately never found a nest. Nevertheless, with over 3,500 freckled ducks inhabiting the lakes, local birdwatchers were privileged to host a considerable proportion of the total population of these birds in the mid-1990s.

Although duck identification came easily, the migratory waders gave me considerable grief. Twice a year mobs of small brown birds drop in on their way to and from Japan, China or Siberia and the southern coast of Australia. These birds are called Palaearctic waders, which in layman's terms means 'bloody hard to identify'. Not only are many roughly the same colour, but most of them change their colour dramatically as they remodel from their colourful northern hemisphere breeding plumage to their drab antipodean garb.

Fortunately Australia now has several excellent field guides with accurate and life-like drawings and detailed descriptions of our birds. These are a considerable improvement on the old *What Bird is That?*, which is an appropriate title given the quality of the illustrations. I used to get frustrated looking through my old man's copy of this ancient tome because many of the honeyeaters, raptors and especially the waders looked simultaneously identical to each other and a far cry from anything I would see in his big old binoculars. Anyway, unlike novice birdos from a generation ago I could no longer use the lack of a suitable field guide as an excuse not to learn the waders.

Invariably two or three skittish species of waders flock together and would spook and fly off before I could work out if their legs were pale green or pale grey, or whether their eyebrow was prominent or not. As with the initial recognition of other types of birds I found that the best way of learning was to get thoroughly acquainted with a few of the common species and compare new ones to these. In this way the red-necked stint was recognisable as a similar size but with a different posture to the familiar red-capped plover, and the curlew sandpiper was like a sharp-tailed sandpiper with a slightly longer, more down-curved bill.

Most of the lakes took a couple of hours to walk around, about the same time it took me to bounce and crash my low-slung Belmont station wagon to get to them along the invariably rocky tracks. When it was full Lake Richardson near Woomera took six hours to circumnavigate. Shell and Coorlay Lagoons and Lake Koolymilka were nearly as big. Often I would try to cut off little bays by walking through the shallow water with my clothes and binoculars on top of my hat. This sometimes proved to be risky in winter if I unintentionally submerged my Plimsoll line into the painfully cold water.

My solo, shore-based counts were generally frustrated by the thousands of ducks drifting over to the opposite side. Picking out a chestnut teal in among a flotilla of grey teal at a distance of a kilometre or more was not a skill that I had acquired. However, the birds were usually within binocular range when they had been forced from the opposite bank by one of my mates whom I coopted to venture out to the lakes. Working with another birdo also kept you on your toes. Scanning each duck is a formidable challenge when we frequently encountered aggregations of over 20,000 birds on a lake. What would be hiding among the masses of hardhead, coots, grey teal and pink-eared ducks that made up the bulk of the counts? With an inventory of over fifty waterfowl species coupled with a seemingly endless list of potential migratory or vagrant species, I was always mindful that a new species for our list might be concealed among the common birds.

I dreaded returning to the car to hear 'How many mountain ducks did you get?' or 'Did you find the great-crested grebe's nest?' if my notebook did not also oblige.

I particularly enjoyed driving down to Lake Koolymilka where the skeletons of the 1950s village and launch pads of the Woomera rocket range were located. A mind-boggling amount of infrastructure lined the pot-holed bitumen road which slashed across the gibber plains. The foundations of the Koolymilka village, which housed hundreds of servicemen, overlooked the lake where swans were breeding in their thousands. In 1976 they even had a yacht club there. I presume it was either a pompous Pom or tongue-in-cheek Aussie who was responsible for the road sign informing motorists that 'kangaroos abound on this road at night'.

Lake Koolymilka was only accessible through locked gates with permission from federal security guards (apparently a squadron of F1 11 fighter jets was stored in one of the huge sheds down there) and it took a lot of persuasion to convince the security team to let me in to count the birds. I didn't want to blow it as I hoped to return and document the changes to the waterbird populations as the lakes dried.

My first 'date' with Katherine Moseby, whom I was destined to see a lot more of in the future, was a wander around Lake Koolymilka on an unseasonably cold December day. We had met through a mutual friend Rachel at the Roxby Tavern at about midnight the previous night. With the bravado afforded by a session of beers and Bundys I called the bluff of the young biologist who had just caught the bus from Adelaide, suggesting that I should pick her up in a few hours' time to look at ducks. At daybreak I had no option but to ignore my fuzzy head and collect my brazen new acquaintance. When she was still smiling after traipsing barefooted for hours over rocks and sticks concealed by freezing mud and swan shit, I got my first inkling of her formidable zest and enthusiasm. Katherine's hair, as much as her looks and spirit, captivated me. Just like the local sand that I loved, the sun played on her hair to reveal a myriad of shades of blonde, orange, copper and red. Later that night I was not the only bee in the Roxby hive to notice the red roses on her miniskirt after she smuggled herself into the annual town Christmas party at the public pool!

Unfortunately attractive young redheads were not the norm among the local bird watchers. My chief accomplices on the lake sorties were usually Fewster, who preferred the shorter local walks to his Lake Eyre ordeal, and Roy Ebdon. Together with his wife Mary, Roy lived and breathed birds. The highlight of their year would invariably be the sighting of a hitherto unrecorded bird or the first nesting for a species that they were able to incorporate into Roy's computerised database. Roy looked somewhat older than his fifty-odd years, and inevitably had a medically treated ailment of some nature. On more than one occasion after he had disappeared from sight on the far side of a lake, I would fear the worst until a few hours later when he would reappear red faced under a white towelling hat, babbling through a grin about his most noteworthy finds. Roy and Mary loved phoning me to gloat about an eastern curlew sighting or nesting plovers. In turn I would taunt them with discoveries of Australian shovelers out of their normal range, or reports of three ibis species on the one lake. Between us we did well over 100 trips to the lakes. We tried to get around as many lakes as possible in a day, so that the overnight movements of waterbirds from lake to lake did not affect our regional tallies. In mid-1992 we conservatively estimated that at least 150,000 waterbirds inhabited the eight remaining lakes. It was an exciting time where each discovery spurred us on to spend more and more time around the lakes, documenting the changes in the birds as the seasons came and went, and as most of the lakes became shallower and saltier.

The arrivals, departures, increases and decreases in the number of different birds varied considerably. Pink-eared ducks bred on the lakes soon after they filled but we did not record any nests in subsequent years. In contrast, the cormorants did not commence breeding for a couple of years after the lakes filled. The most consistent breeders were swans. We found their clutches of about six large white eggs in green waterweed nests almost continuously for six years. The huge but gradual increase in swan numbers to nearly 8,000 was largely attributable to local breeding. By contrast other birds such as ibis and egrets peaked later from influxes but did not breed locally.

Sometimes I would camp out by myself on the shore of the lakes. By sleeping in a swag out on a sand spit I could be in a great position to observe the waders that congregated on the shore overnight. One morning after a sleep punctuated by the continuous movements and sounds of birds, I woke eyeballing an unfamiliar wader. Slowly I pulled out my notebook and rested my binoculars on my pillow. I noted that the wader was built like a plover and was much larger than the red-necked stints that were standing nearby, so much so that they could almost park underneath its wings. The bird had a dark, almost black face and belly, joined by a thin black line down the throat that was prominent against its grey–white breast. The pale eyebrow was relatively distinct. When I propped myself up on my elbows to get a better look, the bird flew off silently, unlike more familiar plovers that call distinctively when flying. In flight it revealed a white rump with five or six dark bands on a white tail.

As I rose, I spooked an assortment of other waders that produced a cacophony of alarm calls led by the loud piping of a greenshank that further disturbed my bird. Eventually I got close enough again to determine that it had a pale wing-bar when it was flying. However, its most distinctive features which were only evident when it flew high were its black 'armpits'. Together these observations were enough for me to identify a grey plover, although my field guide indicated that they were 'uncommon migrants to coastal regions'.

Not only was I excited that I had recorded a 'coastal' migrant far from any coast, but when I learnt of the distance that this bird had travelled I was both amazed and humbled. Grey plovers only breed in frozen arctic tundras and hence this bird had probably flown down from northern Canada, Alaska or Russia to spend a few days at Lake Richardson before continuing its round-the-world flight. Their dazzling endurance and navigational skills would invariably lead them back to their same breeding location. I could not help but wonder what this little grey bird, along with some other unexpected migratory waterbirds such as black-tailed godwits and ruddy turnstones, thought of the tastes, smells and sights of Lake Richardson. They had probably never seen or heard grey teal or hardhead ducks before and now they were surrounded by thousands of them. One can only guess at what the migrant birds made of the woolly merino sheep and red kangaroos that drank at the lake. The last mammals that they co-existed with could well have been polar bears and reindeers.

It was not only the waterbirds that were attracted to the lakes. Large flocks of rainbow bee-eaters, welcome swallows and martins patrolled the skies hawking for insects. Magpie larks and willie wagtails nested in their hundreds in the branches of tea-trees that protruded from the water. During mid-1990 they had to share these trees with over a thousand starlings. Feral birds had never been recorded in such numbers in the region and it was with considerable relief that these flocks of nest-hollow bandits left within a few months of arriving.

Nesting was also frenetic in the hollows of tea-trees that escaped the flooding or had been exposed by dropping water levels. I unintentionally flushed a breeding parrot out of nearly every hollow as I weaved among the tea-trees on the shore of Lake Mary. Bourke's parrots, cockatiels and corellas were nesting in many hollows but it was the budgies that were truly swarming. Bright green squawking arrows burst from every direction. Budgies are particularly randy little birds and usually breed *en masse* whenever button grass is seeding. As I sat in the shade of a tea-tree I watched a pair that had been feeding nearly fledged chicks as they nuzzled up and started necking like doting adolescents. Suddenly in a flurry of green wings and tails they rapidly consummated their passion. One brood a year is not sufficient for budgies that, like zebra finches, breed continuously while conditions are suitable.

Life was not rosy for all the birds on the lakes. Several times the birds that I watched would panic as a predator appeared on the scene.

Roy and I learned to be particularly careful that we did not contribute to predation at nests. One such incident occurred when we were keen to determine how many cormorants were nesting out in the middle of Coorlay Lagoon. As we approached the colony by walking through chest-deep water, many of the adult birds flopped out of their nests and flew off. To our horror several Australian ravens immediately appeared and each made off with a white egg in their beaks. Even after we had hastily retreated, the cormorants made no attempt to defend their nests or chastise the ravens and as a result many eggs were taken. When we investigated a colony of gull-billed terns that was nesting on a sandy island, silver gulls quickly came in and helped themselves to unguarded eggs or chicks. We decided that we would have to monitor these nesting colonies from a distance, which was frustrating but at least not hazardous to the birds.

It was with considerable relief that I could compensate the waterbirds for occasional losses by shooting a couple of feral cats that were hunting at the lakes. Despite the fact that they are supposed to avoid water Roy observed one cat swimming out to a flock of ducks. On another occasion I shot a cat that was stalking some roosting grey teal on the shore. That cat must have developed a taste for duck since its gut contained the partially digested remains of another teal.

Although we also saw several other species of raptors hunting at the lakes, like little falcons and brown goshawks, the predators that drew the most attention were the wedge-tailed eagles. These awesome dark eagles were as thick as flies on some days. Once I counted fourteen in a single thermal over Coorlay Lagoon. Whenever eagles flew over the lakes the ducks bunched up into tight alert groups and grebes dived under water. Occasionally an eagle would swoop into a power dive and snatch up an unwary native hen or duck from the margin of a lake.

I paid particular attention to the wedgies after an incident on a 47°C workday when I used the excuse of getting a water sample from Coorlay Lagoon to jump in and cool off. I left my work-issue brown shorts on the bank and dived in, drinking the cool fresh water as I swam around chasing water beetles. I was a fair way offshore when I noticed a wedgey dive down and pick up my shorts before disappearing over the low cliff face. The loss of my shorts in this manner would have had me laughing had my keys not been in the pocket. I then had to walk across the shimmering hot gibber to sheepishly wait for a car so I could hitch back to Roxby. The worst part of the experience was endlessly relaying the story, as I needed replacement keys for not only the car and my camp accommodation, but also for most of the locked gates in the north of the state.

11

DUCKOFF

Lake Bessie is an out of the way lake on Purple Downs Station, thirty kilometres south of Roxby and rarely visited by locals. It was the sort of place you could walk around all afternoon and not see any footprints or tyre marks: the type of place where there was no rubbish and no fireplaces. At Lake Bessie the birds, their nesting, movements, diets and predators invariably swallowed up my thoughts. At Lake Bessie I got as close as ever to thinking like a duck.

During the day the ducks and swans were largely quiet. What noises they made were generally engulfed by the sound of the wind or the buzzing of the flies. At night when the wind dropped and the flies gave way to the quieter mossies the waterbirds found their voice. One night I rolled my swag out on a beach to better observe the nocturnal antics of the birds and to get a close look at some waders in the morning. The night sky resounded with the calls of thousands of invisible birds moving from lake to lake. I was mesmerised by the symphony played by thousands of swans: the bizarre oboe-like honking of the males, or cobs, accompanied by the squeaky-door sound of the hens. The Lake Bessie swans were ably supported by the laughing of teal, the squeaking of pink-ears and the bleating of wood ducks. These birds, which largely floated around or roosted quietly on the banks by day, were definitely creatures of the night.

To further enhance this great nocturnal masterpiece I found that I could conduct the performance from my swag with the aid of my Maglight. By pointing the torch in their general direction I could increase the gusto of the performance markedly. Thousands of feet clapped against the water in a deafening round of applause as they powered into flight and into the anonymity of darkness. The white wingtips of the swans are only exposed in flight and rippled away from the torch beam like whitecaps on a wave.

Years later, memories of these aquatic symphonies were largely responsible for solving a dilemma that Roxby and most other large mines in Australia were facing. In 1995 the corpses of over 2,000 waterbirds from a cyanide-laced tailings dam in New South Wales were shown across the country on prime time television. Cyanide not only leaches gold from ore bodies but also leaches the life out of birds that come in contact with it. A decade earlier Mt Windarra in Western Australia turned green when 60,000 budgies dying of thirst at the onset of drought perished in a tailings dam. Although these two events were freaks they were enough to make any company using cyanide very nervous and with good reason.

Olympic Dam uses tonnes of cyanide to extract gold and silver from the ore but the metallurgists have worked out how to take the sting out of this valuable and yet incredibly toxic chemical. Most of the waste cyanide is bound up in a stable solid 'goo' called ferrocyanate, and the remainder is rapidly converted to hydrogen cyanide when it mixes with acidic wastes from elsewhere in the plant. Fortunately for the ducks, hydrogen cyanide breaks down in sunlight before it gets to the dams.

Unfortunately cyanide is not the only potential killer in the broth that metallurgists concoct. Although the Olympic Dam tailings ponds are essentially free of cyanide, they are effectively lakes of acid. Unlike cyanide, acidic lakes give birds a sporting chance. Veterinary analyses and repeated sightings of rescued birds suggests that if birds are removed from the acid ponds within a few hours they usually survive. Long-term occupancy typically has dire results.

Like the budgies at Mt Windarra birds will sometimes visit toxic ponds, even if they are of no value to them or even downright hazardous. Waterfowl, in particular, are not very fussy about their selection of landing sites. Thumps on homestead rooves at night or ducks sitting on wet tarmac in the morning, attest to the problem of the waterfowl's tendency to land on some shiny surfaces at night. It is hardly surprising that ducks and swans will occasionally lob onto tailings ponds, especially when they are exhausted after long journeys across the waterless outback. The larger the tailings ponds, the more waterfowl they are likely to attract. In 1995 the size of the ponds at Olympic Dam increased enormously. As a result the waterfowl's use of these ponds increased tenfold. Instead of a dozen or so deaths a year, dozens of birds were dying each month. This was clearly a serious problem.

In the mid-1990s it became apparent that the tailings ponds at Olympic Dam were seeping. When environmental staff did their sums they realised that the amount of liquid being pumped into the ponds as sloppy purple tailings did not equate with the amount of liquid left in the ponds after they had accounted for evaporation. It seems blatantly obvious, in hindsight, that the liquid component of the tailings was seeping through the base of the dam and into the porous rocks below. But at the time it was not so clear-cut.

The ground water level in bores which were monitored to detect leaks around the dams had risen as a result of the phenomenal 1989 rains, and continued high bore levels were thought to be related to the rainwater recharge of the aquifer. Water samples from these bores did not suggest that they were being contaminated by leaking tailings. We now know that as the tailings seeped through the floor of the dams most of the heavy metals and radiation-charged particles were bound up in the clay base. Furthermore, the underlying limestone neutralised the acid so that the inflowing liquid was actually cleaner than the super saline natural ground water in the aquifer. There were no other environmental clues of the leak because the salty aquifer, which is thirty to fifty metres below the ground, does not reach the surface as a spring or lake, nor do trees' roots reach down into it. Rather, the leaked and filtered tailings benignly seeped towards the mine where pumping of the deep drives had created a local sink.

The tailings had seeped because WMC had cut corners. Instead of the putty-like tailings setting like concrete in a series of impervious layers on the base of the dams, the tailings had been kept continuously wet. Therefore, instead of evaporating on top of a purple rock wafer some of the liquid seeped through the mousse. The tailings dam was not contained and the public's outrage rose. More than any other aspect of modern mining, the containment and management of tailings presents the greatest risk to the environment and hence the greatest cause for public and political concern.

It is easy to see why the public were outraged at the prospect of a leaking tailings dam. As the Olympic Dam mine was copping flak, Australia's largest mining company, BHP, was being accused and fined for environmental and social impacts on a huge scale as a result of the tailings from its Ok Tedi mine in Papua New Guinea. Mismanagement of tailings at this and other sites had drawn the public's attention to these blights on the environmental credibility of modern mining operations.

The contrast between Olympic Dam and Ok Tedi could not be more extreme. Perched in the immensely rugged and remote Star Mountains, BHP and the PNG government were essentially removing a mountain of gold and copper, like pulling a steeple from the roof of the tropical nation. In the process millions of tonnes of overburden and waste rock were deposited into the headwaters of the creek from which the mine took it name. These wastes, or tailings, were flushed down the fast-flowing stream towards the mighty Fly River. On the way, as the iced-coffee coloured water slowed, the tailings were deposited on the banks and bed of both the Ok Tedi and Fly rivers, raising the water level and flooding over 1,000 square kilometres of virgin forest.

Ok Tedi is located in a ridiculous, improbable landscape. When I first flew into Ok Tedi my pilot had banked steeply to negotiate the cloud-shrouded airstrip within the rugged valleys and I was struck by the lack of tall trees. I had imagined that a forest that hugged the equator and received ten metres of rainfall a year would support monstrous trees. But the trees could not grow tall because they would simply fall over. Landslides are as much a feature of the Star Mountains as willie-willies are of the Australian outback. The mine site had a team of bulldozers operating continuously just to clear the few local roads of the debris that slides down the precipitous slopes.

Despite several attempts to solve the tailings' problem at Ok Tedi, the mine operators had to concede that they could not construct a safe tailings dam in such a supersaturated and steep landscape. Recently, in response to the environmental and social impacts that have accompanied the mismanagement of tailings, BHP along with some other major mining companies have decreed not to commit to any more mines that dispose of tailings into rivers. While leaky tailings conjure up images like Ok Tedi or the February 2000 Arul Baia Mare mine disaster in Romania and Hungary, the seepage of tailings at Olympic Dam was a completely different story. Even though the tailings had not been managed properly there were no environmental or social impacts whatsoever. This highlights the environmental and engineering advantages of mining in flat, dry deserts as distinct from in tropical mountains or alpine areas.

To 'control' its tailings at Olympic Dam, WMC pumped the liquid component to a series of new plastic-lined evaporation ponds leaving the solid wastes to set hard. As a result, this expensive engineering solution to a perceived environmental issue created a real problem. Suddenly nomadic waterfowl were enticed by a huge series of ponds in the desert. Although the tailings ponds were too acidic for the ducks to drink or to sustain any aquatic life that they could eat, the tenacity to which these birds clung to these stinking ponds at first defied comprehension.

On several mornings I was alerted to a couple of swans or ducks that had taken up residency on the evaporation ponds overnight. Together with as many human scarecrows as I could muster we drove around with blaring horns and wailing sirens, dragging ropes and hurling rocks, paddymelons or firing high powered ammunition as close as possible to the birds. The response was invariably an aloof glance at the source of the disturbance followed by a scarcely perceptible drifting to a more remote position on the pond. Sometimes we were not even afforded the acknowledgement of a response. On a few occasions when we managed to drive them out of the ponds the birds would typically cut a few low circuits and land on an adjacent pond. They were infuriatingly frustrating. All they had to do was to fly to a nearby sewer pond or clean water dam and they would be fine, no more harassment by despondent ecologists and no more exposure to a life-threatening cocktail of chemicals.

Eventually it took a coot to alert me to the reason why we had problems shifting birds from the ponds, and that the solution for the birds wasn't as simple as flying off. This particular black chook-like waterbird had been frustrating me for over an hour. It was on a relatively small pond that was within reach of my shoulder-popping rock throws. I had nearly given up when I did one of those throws that you regret as soon as it leaves your hand. Like a kid in the backyard who has just launched a six towards the window, I tensed-up and twisted my body around in a subconscious attempt to bend the rock away from its target. Shit. Sorry mate, I thought, as the coot was about to be poleaxed by a rock half its size.

I waited for the splash of rock, wings and feet to subside to see what damage I had done. To my amazement, instead of a limp raft of black feathers bobbing to the surface, the coot had evaded the rock and abandoned its stubborn resistance to flying. It gathered the momentum for flight more quickly than I had ever seen, rapidly gaining height. Or at least I thought it was rapidly gaining height. From the corner of my vision appeared another black object, bigger than the coot and that seemed to freeze the bird in time and space. Within seconds there was an explosion of damp feathers. The impact, which is forceful enough to kill an adult kangaroo, obliterated the fleeing bird instantly as the wedge-tailed eagle maintained its trajectory down to a pine tree where it commenced its breakfast. The poor old coot had lost an unwinnable battle. It goes to show that you can't necessarily believe the old adage about a 'sitting duck'. With the exception of the first few seconds of open season, sitting ducks are far safer than flying ducks.

I subsequently witnessed little eagles and peregrine falcons repeating the treatment to waterfowl that we had flushed from other ponds. Clearly the reason for these waterbirds to stay on the ponds during the day and to fly at night was to avoid predation. Fear of falcons or eagles is also one of the main reasons that most huge bird migrations, which may span entire continents, usually occur at night. If the ducks only flew at night I needed to shift them from the ponds under the cover of darkness. Flags, wires, kites and other 'scarecrows' which were visible by day were not going to work, but what would? Then I remembered my lakeside symphonies. I would move the birds with lights.

※ ※ ※

The Roxby Downs sewer ponds were an ideal location to trial a range of different lights to establish which ones the ducks found least tolerable. All over Australia and particularly in dry areas, sewer ponds provide a somewhat odorous refuge for waterfowl and many other birds. Whereas bird visits to the tailings ponds were both infrequent and low in numbers, I was guaranteed of having plenty of birds at the sewer ponds to try out various forms of nocturnal harassment. If a torch could rouse ducks and swans from a huge lake, I figured that flashing lights might scare them from a small pond into another postcode.

Just on dusk I crept around the ponds through clouds of midges and deposited a car battery with a light attached to it on the bank. By walking to the opposite side of the ponds I could ensure that most of the ducks would be congregated near the light, which was switched off. Just after dark I would then sneak around and connect the light and watch the response of the birds. The first one I tried was an orange rotating beacon like the ones used by road workers and mine vehicles. To my dismay the birds paid little attention to the light and definitely did not panic *en masse* as I had hoped. I actually got a bigger shock than the birds that night when the dormant inflow pipe suddenly geysered into action as if half the township had simultaneously flushed their dunnies. The fact that the birds didn't respond to this sudden and noisy gush of water concerned me. Maybe they would quickly habitualise to any deterrent that I tried?

The following night I repeated the exercise, this time using a brilliant white strobe. An ecologist ordering a strobe light raised a few eyebrows from the mine store and even more so when I checked it out in the laboratory. Staring at the light from close range was almost impossible and even walking in its vicinity was difficult in daylight, due to the distortions of the rapidly flashing light. I figured that this would have to freak the ducks out. I had visions of erecting dozens of strobes around the tailings dam and maybe a few gigantic mirror balls for good effect. However, the only action that I could generate from the ducks that night was a couple of surprised grunts and squeaks as I shielded my eyes from the strobe. I could not believe it. My plans for a nocturnal deterrent appeared to be flawed.

The next night I made one last attempt and got back to basics with a spotlight. When the light was switched on all of the birds illuminated by the beam immediately took to the air in a flurry of wings and squawks. Although some ducks outside the light were also swept up in the panic, it was apparent that many of the birds in the dark were not affected. I quickly discovered I could move all of the birds from a pond by 'painting' the water surface with the spotlight beam.

The higher I stood with the light, the smaller and rounder was the beam on the water. It did not take long to confirm that the largest spotlight beam was created when the light was held close to the water. Because the level of these ponds rise and fall the rotating spotlight would best be mounted on a raft, and because it was on a raft the light should be solar powered. One other specification concerned me. As anyone who has driven or spotlighted in the country could attest, moths and other insects are attracted to lights like flies to a sore. To prevent attracting swarms of insects, which would in turn attract bats and birds, the light should turn on and off intermittently. With the assistance of some electrical staff DUCKOFF was born.

I was satisfied that my theories were sound but still not confident that DUCKOFF would have the desired effect after the initial shock of the light wore off. So I took the prototype mounted on a raft to the sewer ponds for some trials. True to specifications the timer started the first rotation just after dark and scattered the ducks. Every minute or so it would flash back on for a couple of rotations to scare off any remaining birds or any that had subsequently landed on the dark pond. At dawn the next morning I was relieved to see that my floating light was in the middle of a barren, duckless pond. Over 100 ducks had crowded onto the ponds without DUCKOFF. One night is hardly conclusive proof so I went back the following dawn. To my dismay the ducks were spread relatively evenly over all of the ponds. What's more the raft was nowhere to be seen. I couldn't believe it, the bloody thing had sunk! There was nothing to do but strip off and walk out into the middle of the slimy, stinky pond and retrieve my prized new invention. The kookaburra is widely acclaimed as having the most recognisable 'laugh' of any Australian bird, but grey teal are actually the masters of this art. As I heaved the raft onto the bank and allowed the cold green soup to pour from its flooded chambers, I swear the teal were deliberately goading me with their derisive chortling. But eventually I had the last laugh. I got the raft repaired and moved it from pond to pond. Each morning the ducks, including the smug teal, were congregated where the DUCKOFF wasn't.

Further trials showed that DUCKOFF effectively deterred ninety percent of the ducks. Some of the birds that were not displaced by the rotating beacon were either sick or moulting and hence were unable to fly away. Together with gas guns that emitted loud bangs at irregular intervals the DUCKOFF lights significantly reduced the numbers of waterfowl recorded on the tailings. Even more rewarding is that other mines have adopted similar strategies to help reduce waterfowl deaths. The poor coot that had survived a bombing only to be wiped out by a wedgey had not died in vain.

Since the mid-1990s most of the local lakes have been dry and the vast majority of the waterbirds have moved on. All that remains at the lakes is a new high-tide line of young tea-trees and a smattering of eggshells. With the departure of most of the waterbirds their exposure to the Olympic Dam's tailings dams has decreased considerably. It is difficult to judge the continued effectiveness of the DUCKOFF system when there are not many birds around. But they will be back sometime. When they do return a bank of rotating beacons rather than red-faced, rock-throwing biologists will confront them.

PART THREE

Scales and Tails

12

Life on the moon

The number twenty-three is instantly recognisable to sports fans around the world. Michael Jordan, perhaps the world's most gifted and richest sportsmen of all time, wore '23'. So does Shane Warne when he flips out his leg spin in the pastels of one-day cricket. Consecutive premierships for the Adelaide Crows were won largely through the skills of a young champion Andrew McLeod who donned the same magic number. Twenty-three is also the Holy Grail for many Australian herpetologists.

One of the first things that Aussie snake lovers learn is the significance of the number twenty-three. Many snakes can be incredibly difficult to identify by their colour or shape alone and only experienced herpetologists can avoid inspecting a snake's scales to establish its true identity. Most of our large venomous snakes have seventeen or nineteen rows of scales around the middle of their body. Browns, dugites, mulga snakes and red-bellied black snakes all fall into this category. The exceptions are those with fifteen scale rows or those snakes that really get the heart pumping—those with twenty-one or twenty-three scale rows.

When you get past nineteen on a scale row count of a serpent flexing in your hands you know you have reached the big time. In all probability you have just landed a taipan, the most feared snake in Australia. Reaching lengths in excess of two and a half metres and armed with venom nearly eight times as toxic as the notorious Indian cobra, there is ample reason for this status. Until the 1980s the taipan held the title of the most venomous snake in the world.

In the late 1960s and early 1970s a number of large brown snakes turned up in the Birdsville region of south-west Queensland. Jeanette Covacevich and her co-workers from the Queensland Museum were able to show that these snakes were the same as two specimens probably collected near the confluence of the Murray and Darling rivers in 1879. Rediscovery of reptiles usually creates a bit of excitement in the herpetology world but such discoveries typically leave the front pages of the newspapers pretty quickly. However these Birdsville snakes were different. It didn't take long for Covacevich to realise that these elusive snakes, with twenty-three mid-body rows of scales, were related to the taipan. A rash of studies into their ecology and toxicity soon followed. As a result the much revered 'original' taipan, subsequently known as the eastern or coastal taipan, was dethroned as kingpin of venomous snakes.

Straun Sutherland, the architect behind the development of antivenom for many of Australia's snakes, showed that the 'new' taipan had frighteningly toxic venom. The standard measure for venom toxicity is the amount required to kill half of the laboratory mice that are injected. This new snake, which was variously known as the inland taipan, western taipan, small-scaled snake or fierce snake, had the most toxic snake venom ever recorded. But toxicity is not everything when it comes to killing mice, or anything else for that matter. The amount of venom injected in a bite also plays a role. This is when the inland taipan gets really scary: one bite from an inland taipan can contain sufficient venom to kill over 500,000 mice.

Half a million mice would be enough to feed 2,000 generations of taipans. Why on Earth would a snake need to be so venomous? The eastern diamond back rattlesnake, infamous for killing American cowboys within seconds in Hollywood westerns, is 100 times less venomous, yet they seem to get by ok. Inland taipan venom seems to be the very epitome of overkill. Or is it?

Out in the channel country of the Cooper and Diamantina rivers more inland taipans were discovered as herpetologists rushed to study these incredible snakes. They live on cracking clay plains and like their coastal cousins the inland taipan feeds almost exclusively on rats. The native rat of the channel country is the long-haired or plague rat. Stockmen dreaded camping among plagues of rats that chewed everything in their path including rations, saddles, swags and even tyres. The first time I ever saw one of these infamous rats was wandering back with some mates from the Birdsville Pub to our campsite. We startled a rat from the luxuriant growth in the Diamantina floodplain and I decided to check out this creature that I had heard so much about. I dived at the rat and immediately accumulated a dense mat of bogan flea prickles. These most persistent prickles break into a multitude of parts and take forever to remove. My second dive yielded the feisty rodent that squeaked and squirmed and chomped into my fingers. Holding the long-haired rat for a few seconds made me acutely aware of why the inland taipan has such toxic venom.

A rat can easily kill a snake. The sharp teeth that were gnashing away at my hands can remove a snake's eye or even pierce its skull. I am not the only person who has endured the surprise and shame of keeping a large venomous snake that has been mortally wounded by a mouse that was supposed to be its dinner. If a mere house mouse could kill a large snake, a taipan has good reason to fear one of these long-toothed hairy monsters.

In order to prevent their prey from escaping or inflicting serious bites, brown snakes will often coil around a mouse, python style, while waiting for the venom to take effect. A snake would not risk this strategy with a long-haired power-pack. Instead, inland taipans rapidly strike and release their prey. Weaker venom could still kill a bitten rat but it may scamper off and die out of range. The stronger the venom the better chance the inland taipan has of converting a potentially dangerous rat into a meal. From the benign saliva of its ancestors, natural selection has shaped the evolution of this cocktail of paralysing neurotoxins, muscle-destroying myotoxins, blood cell destroying haemotoxins, anticoagulants and coagulants, into an essential component of the inland taipan's ecology.

※ ※ ※

The Birdsville Pub was not the only northern pub where interesting biological discoveries were made. I had left jars of laboratory-grade alcohol at several outback pubs and roadhouses for the custodians to preserve any strange critters who were the victim of roadkill, the family pool, moggie or shotgun. My most successful jar was at the William Creek Pub, probably due to my frequent visits to this rustic watering hole. William Creek is a couple of hours drive from any town and about 270 kilometres from Roxby, a handy drive to build up a thirst on a Friday night. Such is the magnetism of the pub, that William Creek must be the only town in Australia where most of the residents have managed the pub at some stage. Across the road is a newly established café that challenges the famous Oodnaburgers for the best take away food in the region. It was therefore a convenient launching spot for weekend expeditions to Dalhousie Springs, Lake Cadibarrawirracanna, the Denison Ranges or Coober Pedy. As a result, the William Creek Hotel was almost my 'local' and together with a carload of mates I spent many good nights yarning with the Anna Creek ringers, and leading astray the overseas tourists who converged there.

In April 1992 a local fencing contractor by the name of Jeff Boland rang from the William Creek Pub to tell me that he had collected a snake from the Coober Pedy road that looked like a taipan. He said he would drop it in my jar at the pub. Inland taipans were restricted to the channel country, over 500 kilometres away, so I suspected that Jeff had collected a western brown snake, which comes in a wide range of sizes and colours. Nevertheless, a mate from Roxby who was propping up the bar retrieved the jar for me.

A few days later I went through the contents that included a couple of snakes, lizards and insects. The first snake I pulled out was different to the browns that I normally found in the jars. It was a blue–grey colour with a yellowish tail and only fifteen scale rows. At only 15.5 grams, but over half a metre in length, it was a slender strip of a snake and the white markings around the eye confirmed that this little fella was a desert whipsnake. I only rarely came across these colourful little speedsters, so I was pretty chuffed as I recorded its details. The second snake was badly mangled but looked like a brown as I held it up and let the alcohol and reptile gravy drip back into the jar. Instinctively I did a scale count. Twenty-three.

Twenty-three! I must have got it wrong. Sometimes an injury or glitch in the scalation caused scale counts to deviate from the norm on a particular cross-section of the body. I counted again, and again, and again, up and down the length of the snake. Twenty-three every time. I looked at the head. It had the long, coffin-shaped appearance typical of taipans. From the tip of the nose to the 'vent', which is the lingo for a reptile's private parts at the end of the body, it measured approximately 1,265 millimetres, the tail adding a further 220 millimetres. The anal scale at the base of the tail was single, not divided as in some other snakes. Oh my goodness!

Even though it was long dead I went all sweaty as I realised that I was holding the most venomous snake in the world. Gingerly I prized open the mouth with forceps to look at the infamous artillery. Nestled imperceptibly within their sheaths of skin was proof of the old adage that size doesn't count.

I raced outside and took some photos of the mangled corpse then looked up my snake books. Coober Pedy was approximately 500 kilometres from the nearest confirmed inland taipan localities, up near Birdsville and Moomba in the far north-east of the State. This find was clearly something special, but I didn't know how special. First I had to confirm the identity of the snake. Together with Mark Hutchinson, curator of reptiles at the South Australian Museum, I confirmed that Jeff's snake closely matched museum collections from the Moomba region. It was a dinky-di inland taipan and not a hitherto undescribed species.

The next question, which was of more interest to me, was how did it get to Coober Pedy? Coober was a large town by outback standards and had been occupied since 1915, so it seemed unlikely that a large snake that lived in the district could escape detection for so long. Chances were that it fell out of a stock truck or tourist vehicle that had driven down from the channel country. I wanted to get to the bottom of the mystery of the snake's provenance. Even though finding another snake in the district was a long shot, it was the only way of determining whether Jeff's snake was a resident or a hitchhiker.

The next weekend while driving towards Coober with Fewster and another friend Marilyn Holden, I was looking out for potential inland taipan habitats. Its channel country haunts are flat clay plains bisected by eucalypt-lined creeks. These creeks which flowed most years flooded into swamps vegetated by wiry lignum and canegrass. I was hoping to find swamps like these that might harbour rats and taipans. Long-haired rat country would have to be wetter and better vegetated than the endless dry gibber plains and the low orange and yellow sandhills that I was driving through. The creeks draining into Lake Cadibarrawirracanna looked like my best bet. Before I started poking around these creeks I wanted to find out from Jeff Boland exactly where the snake had been collected.

Coober Pedy, like Andamooka and most caravan parks, was an impossible place to locate anyone's abode. Not only were most streets unnamed, the houses weren't even numbered. In fact at Coober, like most opal towns, it is sometimes difficult to determine what actually constitutes a house. Wheel-less caravans and buses sometimes house opal miners, while expansive underground residences hide behind a simple door at the base of a hill.

Through a few phone calls I tracked down Jeff at the Italian club—a large building near the main drag. Jeff's thick-fingered handshake bore testimony to the hundreds of stockyards and fences he had built in the bush, some before I was born. He told me that the taipan was found about four kilometres east of the dingo fence on the Coober Pedy–William Creek Road; a long way from the creeks that I had thought looked promising. This stretch of road was only distinguishable from the surrounding barren plains by the graded row of rocks on the road verge. Surely the taipan was an alien in this sort of landscape. Finding a new population had been a pipedream. It looked like we would be spending the remainder of the afternoon yarning at the Italian club about the other critters that Jeff and his drinking partners had seen around Coober. Then, under his manicured white moustache, Jeff casually mentioned that he had recently found a snake skin also with twenty-three scale rows. He had picked it up while checking the Dog Fence on the road out to Oodnadatta. This was too far away to belong to the dead taipan. Surely two cross-country hitchhikers in the same month was too much of a coincidence. Maybe there is a population of taipans living out there, I thought. Buoyed with this new information Fewster, Marilyn and I rushed out of the airconditioning into the hot dusty street. We were going to find an inland taipan.

My enthusiasm rapidly waned when we visited both sites. The first was on the William Creek Road and was a most unremarkable piece of country. The other site was worse! The Dog Fence crosses the Coober Pedy–Oodnadatta Road on the desolate and appropriately named Moon Plain. As I looked out the window the horizon was huge but I could not see a single plant. This country was used for the extraterrestrial-looking shots from Mad Max. Patches of gypsum shards glistened like a thousand broken windscreens on the grey puffy soil. In some areas fist-sized rocks were scattered over the barren ground. It could have been the same location from where NASA's Sojourner Rover transmitted its amazing photos of Martian rocks with names like Yogi, Bam Bam and Pooh Bear in 1999.

The Moon Plain was certainly not prime country for long-haired rats. Like the perenties that used to be seen near Port Augusta these taipans probably represented misplaced individuals that were travellers on some outback adventure. They had ended up way out of their range in a hostile environment. We dropped our dacks and took an obligatory photo next to the MOON PLAIN sign and drove back to Roxby, disappointed that we had not made a great discovery.

Incredibly a few months later Jeff rang and said he had seen another taipan on the Moon Plain. Like the first it was also heading west and Jeff figured that they were moving towards Coober Pedy from their channel country domain. It was possible that long-haired rats had spread through the country and that the occasional inland taipan had followed. Old George Bell from Dulkaninna, on the other side of Lake Eyre, told me he had witnessed two plagues of long-haired rats and he reckoned that they got the occasional 'fierce snake' down there as well. Jeff confirmed that long-haired rats had terrorised the Coober Pedy region for a brief period in the mid-1970s but had not been seen since. I had to get back up to the Moon Plain before any rats left.

The next weekend was the Curdimurka Ball, a black tie show at the old railway siding attended by several thousand revellers. Together with a few mates we had organised the inaugural Hair of the Dog Café for the next day where we fed, watered and entertained hungover ball goers to raise money for the Flying Doctor. The Moon Plain visit had to wait another fortnight until the October long weekend, which was also the Coober Pedy race weekend, a great time to do some trapping and to catch up with the locals. By sheer coincidence a mate from uni, Crowie, was also there. He was working for a Melbourne-based mining company and poking around on the Moon Plain for evidence of hydrocarbons. It was amazing that we were both searching for elusive yet very different riches in the same remote and inhospitable place.

Mick Evans, who had accompanied me on several field trips to remote places, brought his three young sons along. Known even by their parents as 'the ferals', Ben, Connor and Christo spent the entire weekend charging off across the plain exploring, fighting and only reappearing when Mick whistled them back like sheepdogs. Another mate called Staggs, who was keen to see a taipan from a safe distance, also came and helped to set traps at the sites where inland taipans had been recorded. Although the inland taipans had provided the inspiration for our trip I was not really expecting to catch a snake. Rather we were trying to find out whether there were any long-haired rats or other suitable prey mammals for them in this habitat.

Even while we were locating and digging the sites I sensed that the Moon Plain was special. At a couple of the areas we flushed pairs of flock bronzewings, a rare species that was seldom recorded this far south. Flock bronzewings used to be super-abundant nomads with flocks that numbered in the millions, as recorded by the ornithologist LR Reese in 1924. Once again, the past tense is imperative when describing the abundance of desert fauna. Flocks like Reese described have not been witnessed since domestic stock and rabbits razed the vegetation from most of the desert plains. Being ground-nesters, flock bronzewings are particularly susceptible to predation by foxes and cats.

The demise of the flock bronzewing bears a haunting resemblance to the story of the passenger pigeon of North America. The passenger pigeon used to travel in huge flocks which blackened the sky. In the 1800s flocks of over two billion birds were counted. Yeah, that's right, two thousand million birds. By 1900 they were extinct in the wild. Flocking birds, like herding mammals, schooling fish and swarming insects, form dense aggregations because it helps them to find patchily distributed food and affords them enhanced protection from predators. Passenger pigeon flocks were hunted to below the size at which they operated efficiently and as a result the species, which was so successful in huge numbers was rapidly relegated to the genetic scrap heap.

Flock bronzewings were not our only exciting discovery as we set traps on the Moon Plain. Over the next couple of days we also found a dozen gibber dragons basking on rocks on the side of the road. Twelve lizards does not sound like everyone's idea of a big weekend but I was stoked. Gibber dragons were no ordinary lizards. They were only described by scientists in 1974 and had only been recorded from a spattering of sites in an arc between Marree and Oodnadatta. This region was my stomping ground but I had previously only turned up a couple of these enigmatic lizards.

It turned out to be an amazing weekend for reptile discoveries. As well as the gibber dragons we found a slider lizard that I couldn't identify among the leaf-litter that had accumulated under the Dog Fence. Subsequently I learnt this small brown skink with tiny vestigial legs was a new species called *Lerista elongata* that had only been described a few months earlier. Mine was the seventh individual to be taken into the museum. If I had been chasing taipans a few years earlier I may well have 'discovered' this new species. We also found the desiccated remains of another inland taipan.

As good as they were, reptiles did not take the points on our weekend on the Moon Plain. The Australian arid zone is renowned for having the most diverse lizard assemblages in the world. It was not altogether surprising that we would turn up a number of rare lizards in a weekend, especially in an area like the Moon Plain that had seldom been surveyed. What was outstanding about our weekend was the number of mammals that we found.

Our traps yielded planigales, which are ferocious six gram carnivorous midgets who tear apart and consume spiders that dwarf them. Although they look like small flattened mice, planigales are in fact marsupials, like many Australian mammals. A couple of species of dunnarts, the planigale's larger cousin, also peered up at us from our traps. Some of these baggy-furred insect and spider eaters had their pouches full of tiny pink jellybean babies. We also turned up Forrest's mice, which look like a large house mouse with a short tail. Similar to the long-haired rats that we were chasing, Forrest's mice are rodents. Trap checkers quickly learn to distinguish these small-eared mice with large angular heads from their smaller feral cousins after they sink their long yellow teeth into exposed flesh.

The *pièce de résistance* on the Moon Plain that morning was a critter larger than a Forrest's mouse, but smaller than a long-haired rat. It was also a far more docile creature and one which in hindsight may have been the key to the inland taipans' presence in this region. Two months earlier I would not have recognised the mammal in our trap; one second before I would not have believed that they lived out here. Plains rats were a threatened species that had been recorded very infrequently in recent years. They were occasionally found in gilgai depressions in Oodnadatta saltbush country, in the far north of South Australia, and that was the only place I had ever seen one.

Six weeks earlier I was assisting with reptile studies for an expedition near Dalhousie Springs, which is adjacent to the Northern Territory border. My counterpart 'doing' the mammals was Helen Owens. Hels was infectiously likeable and her dry sense of humour and enthusiasm rubbed off on all of the volunteers who were assigned to the reptile and mammal group. One night when we were spotlighting one of the volunteers thought she saw a guinea pig. A guinea pig! Out in the desert! Although we all cracked up laughing at the improbability of the sighting her story aroused our curiosity.

A few days later the mystery was resolved. To our amazement in one of our pit traps we had caught a fluffy little mammal which when sitting on its tail bore a distinct resemblance to a small guinea pig! Hels and I were stoked. Our plains rat was docile and made no attempt to bite when handled, unlike the long-haired rat from the other side of the Simpson Desert. A more appropriate name would have been the 'plains cutie', but zoological nomenclature is not so adventurous.

Coincidentally the reptile curator Mark Hutchinson was returning to the museum from the same field trip. He decided to drive back to Adelaide via the Oodnadatta–Coober Pedy Road to check out the Moon Plain. Slap bang in the middle of the Moon Plain he came across a recently killed, large dark snake. A quick scale count confirmed that he had found the fourth inland taipan near Coober. The site where I was looking at my second plains rat was adjacent to where Mark had collected his taipan. The little cutie was over 100-kilometres south of any locality where it had been found in recent years.

Armed with this new knowledge of the seemingly barren and hostile plain we beelined back to the races to catch up with the locals and share our newfound discoveries. Several of the old-timers confirmed that long-haired rats had made it down to Coober Pedy in the mid-1970s but no-one I spoke with had ever seen a plains rat. Tony Williams from Mt Barry Station, which includes much of the Moon Plain, considered that both taipans and plains rats were 'as rare as rocking-horse shit'. Jeff Boland was pleased that we had found another taipan and were starting to piece the puzzle together.

I was keen to establish where else plains rats and inland taipans might live. My only clue was the cracking gypseous clays of the Moon Plain. My greatest assets were the pastoralists who knew the country back to front. By standing just outside the galvanised iron bar I intercepted a stream of informants.

The first bloke who walked up wore three times as much clothing as I could bear to wear on a warm spring day. While I was hot in shorts and a tee-shirt, Bobby Hunter, my story-telling mate, was wearing a couple of shirts including a thick long-sleeved 'flanny' and his characteristic necktie tied loosely round his neck. Bobby suggested that I should check around Pup's Lagoon and Wirringilpinna on Stuart's Creek, as these areas had cracking clay plains. Bob, like most of the other blokes, was more than a bit interested in the fact that this super-venomous snake might live in his country.

Keith Greenfield arrived next, which was unexpected because along with his wife Lorraine he usually busied himself with the organisation of race meets during the day and only propped up the bar after dark. I asked Keith if he had any really shitty country on Billa kalina Station, for want of a better description of the gypseous plains that must be of limited use for cattle. Such an inquiry is usually an insult to a pastoralist, especially coming from someone with green inclinations. 'That sounds like all of my country', he joked and then fixed me with a more serious eye and said, 'Now why would you want to know about that?' Informing pastoralists that they may have rare animal species on their lease that could interfere with their cattle management would make some people reluctant to talk, but Keith was more intrigued than concerned by my question. He told me of an area on Billa kalina called the Dismal Plain, which had been named for very good reason. It was devoid of perennial vegetation and was so deeply cracked that cattle, horses and even people had difficulty walking on it. Peter Langdon, who used to work on Billa kalina and who is a keen naturalist and talented artist, later confirmed that the Dismal Plain and Bakewell Bore areas would be worth a look.

Next weekend I checked out these areas with a workmate Steve Green, his wife Tan and Fewster. The Greenfields at the homestead were amazed that we were actually going out to look for inland taipans. On the way out to Dismal Plain we again saw flock bronzewings, which I interpreted as a sign that the country was in good nick. We also disturbed a flock of letter-winged kites that were probably ready to leave for their evening's hunting. Like the taipan, these beautiful ruby-eyed raptors feed on long-haired rats in the channel country, so I assumed that they may have located some plains rats on the Dismal Plain. The anticipation was growing. In the centre of the plain we found some huge cracks and could not dig the pit traps in quickly enough. The rains had invigorated a flush of green growth and with the anticipation of more plains rats, and possibly a taipan, we considered that this flat expanse rimmed by low flat-topped breakaways was anything but dismal.

Sure enough the next morning we scored a jackpot. We extended the recent southerly range for plains rats by a further 100 kilometres and bagged another planigale. However, the significance of these two small mammals was lost on Tan who exclaimed, 'Is that all you got?', half expecting us to return with our arms full of taipans.

The Bakewell Bore region proved to be not so suitable for plains rats and our traps were empty, but we still made another interesting discovery. After being plagued by ants the previous night near the Dismal Plain, we decided to camp on a barren claypan adjacent to the bore. Fewster was collecting some firewood when a small gecko leaped off the wood and wobbled under a bush. He called me over and we caught the little fella in the low scraggly bush that resembled a cross between a bluebush and a samphire. The gecko was a tree dtella, one of the most widespread and common geckos in arid Australia, particularly around houses where they congregate to feed on moths. It was not the gecko that took my eye but the bush we retrieved it from.

The bush was a rare plant called *Hemichroa mesembryanthema* and we had just discovered the third population known in South Australia. Frank Badman, my botanist workmate, had rediscovered this plant at Strangways Springs in 1984 after it had not been recorded for 111 years since the explorer Ernest Giles found it near Lake Eyre. After Frank had first shown me the plant a year earlier I had found a population near the Willparoona Springs south of Oodnadatta, coincidentally while chasing another lizard.

If we had found a live taipan early in our quest I may have missed out on many of the stories of these fascinating cracking plains. Subsequent weekends yielded more gibber dragons, Forrest's mice, dunnarts, planigales and letter-winged kites near Pup's Lagoon, Wirringilpinna, Curdimurka and Coward Cliff. This was all very interesting but I was becoming impatient. I had found plenty of bait but I still had not landed a live inland taipan. I was intent on witnessing this incredibly venomous yet apparently secretive snake in action. Would they live up to their name of the fierce snake? From what I had seen, the Moon Plain seemed to be the largest and best-developed patch of taipan country. Although it was a few hours drive further away from the sites I was visiting, I knew that my best chance was to go back to where all of the inland taipans had been found.

13
Taipan country

I had been spending a fair bit of time with Katherine, who had thrived on her walk around Lake Koolymilka and who was keen to come back up from Adelaide whenever there was a party or a field trip on. She was initially not too sure about chasing venomous snakes but quickly became familiar with the variety of mammals and birds that we encountered. With an assortment of other mates we would usually pack the esky and swags and slip away from work a bit early on a Friday afternoon. The girls had discovered Strongbow White, which was bottled venom that masqueraded as cider. After setting pit and Elliott traps and polishing off a camp-oven rabbit stew we would go 'spotlighting' on the Moon Plain. The spotlight actually spent most of the time shining in the esky or flashing around the inky sky, while the errant spotlighter was too busy with a joke to concentrate on the search for 'guinea pigs' or snakes.

In five more trapping visits to the Moon Plain in the autumn and winter of 1993 we caught many more mammals and learnt a lot about this unique landscape. Rains in late 1992 had inspired the frenetic growth of short-lived vegetation that totally changed the sight, smell and feel of the plain. However, due to the properties of the gypseous sediments, soil moisture was quickly sapped and the plain reverted to its grey puffy norm. Because the swelling and shrinking gypseous soils move so much, the roots of some perennial plants are effectively sheared off, which accounts in part for the scarcity of long-lived bushes or trees.

Another striking feature of trapping in taipan country was the scarcity of ants and termites. A couple of times we got caught in summer rain that prevented us from leaving because the roads turned to soup and the soil to glue. In most arid localities winged termites and ants start swarming at such times but we did not witness that here. Clearly the Moon Plain was different to most Australian deserts. Perhaps the rapidly drying and cracking soil is not conducive for colonial insects that construct intricate tunnels and galleries. Instead of termites the main detritivores seemed to be crickets that lived in the cracks. The abundance of crickets provided a great food source for the insectivorous mammals that we frequently trapped. By contrast, lizards were scarce out on the plain because they are more efficient predators of termites and ants.

While I was concentrating my efforts on the Moon Plain, Robert Brandle from the National Parks and Wildlife Service greatly expanded our Dismal Plain trap site. Robert had commenced a study on plains rats and over the next seven years, 466 of these 'rare' rodents were captured at the trapping grid on the Dismal Plain. Unlike many other arid zone mammals that roam widely the rats were relatively sedentary. Animal captures fluctuated markedly from over sixty caught in four nights down to none on subsequent visits, when the plain reverted to its more typical barren and seemingly lifeless norm. After rain, however, the populations always bounced back. Katherine, who was rapidly becoming an authority on outback rodents, then Helen Owens, who took over trapping this site from Robert, found that the rats favoured the deeply cracked sections of the landscape particularly when conditions got tougher.

These deep-cracking soils are like an absorbent towel. If they get really wet the 'towel' gets saturated, the vegetation explodes out of the once barren plain and there is not a better place in the world for plains rats to breed. Robert and Katherine suggest that sixty millimetres is enough rain to 'soak the towel' and stimulate a mass breeding in plains rats. However, storms of this magnitude do not occur every year. In fact these cracking-clay 'towels' usually only get soaked a couple of times a decade.

At another of their study sites the cracked puffy soil occurred in patches among stone-strewn gibber. Here they estimated that only twenty millimetres of rain was enough to elicit a response from the vegetation in the areas where the gibber stones form an impervious pavement. Rather than acting like a towel and catching water that falls on to it, the patches of cracking clay within the gibber pavement rapidly accumulate water from a much larger catchment.

Boffins have a term for this concentration of water into small patches of the landscape. They say that these patches have a high 'rainfall-use efficiency'. These small patches generate considerable growth from little input; they get maximum bang for their buck. Because lighter showers occur more frequently than massive 'towel soakers', rats and other mammals have more opportunities to breed and maintain their populations in the patchy clay than in the larger 'towel' habitats. Therefore the rats can hedge their bets by inhabiting areas close to both the small patches which benefit from light rains, and the large clay plains that provide an ideal habitat after the big rains.

The Moon Plain has a similar arrangement of landforms on a larger scale. Rain washes straight off the surrounding gibber pavements and especially the Breakaway Ranges, and is fed onto the plain through a network of creeks. These surrounding gibber plains and ranges are impervious and thus improve the rainfall-use efficiency of key parts of the plain. Rather than being a flat homogenous area the Moon Plain has extensive rain-collecting habitats that explain why it is a prime location for plains rats and inland taipans.

By now we had discovered much of the ecology of the Moon Plain but I had still not seen a live taipan. One morning my luck changed. Steve Green, who accompanied me on several early taipan searches was now the second environmental manager at Roxby since Barry had left. Like me, Steve felt that sometimes you had to follow your instincts rather than the rules. That is why we drove to Coober one work day. All of my other taipan quests had been squeezed into weekends. Jeff Boland had rung the previous night and said that there was a taipan ten metres down an opal shaft. Work would have to wait.

We sped up the Stuart Highway, slipped through the rain among the white cones of opal shaft spoils and past the Coober Pedy airport. The shaft was only about one metre in diameter and water and mud flowed in as I peered over the edge. Sure enough, just distinguishable in the dull glow of my head torch was a large dark snake. I stuffed a snake bag and some bandages in my pockets and pulled up my socks. Gingerly I lowered myself over the edge and started climbing down the hanging ladder. Normally descending these makeshift ladders dislodges gravelly sand, but now, in the rain I was inadvertently pelting the poor snake with dumplings of mud.

When catching a snake it is handy to have a bit of room to move, to be able to jump backwards if the need arises. One-metre diameter shafts do not afford such luxuries, so in these cases you have to pin the snake from above while hanging off the ladder with one arm. The ladder was slippery and the snake was concealed in shadow. I did not want to get within striking distance so the capture was a challenge. Whenever Steve and Jeff looked in from above they dislodged splats of mud that compounded my problems. The ladder was wobbling in time with my shaking knees. Only after I had pinned the snake and had a close inspection did I realise that our thousand-kilometre odyssey was for a mulga snake.

The trip wasn't totally wasted as we caught up with Barry Lewis who had recently shot a taipan at his dugout on Tom Cat Hill. I found it amazing that they were now turning up in town when they were not even known in the district just over a year ago. Six months later I retrieved another mulga snake from a shaft at Coober Pedy. It looked like I would never get my prize.

Some things take priority over the endless quest for elusive snakes and footy finals are one of them. I decided to take a break from taipaning when the Adelaide Crows made it to their first finals. The drive to Melbourne, the weekend of catching up with ex-Roxby mates and my first visit to the hallowed MCG, where we drank enough to be oblivious of our side's last half fade out, took its toll. On my return to Roxby, somewhat bedraggled after this big weekend, I was given a foam broccoli box with a clear perspex lid that had my name on it. Inside was a present even better than a Crows' Grand Final berth. A large black and very pissed-off snake was glaring at me. Jeff had caught his first ever snake—an adult inland taipan. Somehow he managed to get it off the Moon Plain and into a mailbag, then out of the bag and into broccoli box with a makeshift lid. It was almost an anticlimax after all the hours I had put in. Jeff had found the snake next to the Dog Fence, not far from where we had installed one of our lines of traps. It had been a wet windy day with a maximum of only 15°C, not exactly what is considered to be prime snake weather.

Carefully I peeled off the masking tape and tried to get my fuzzy brain around the fact that I was about to embark on the biggest event of my snake-handling career. Half a million mice, I thought. When I picked it up the snake turned out to have a white belly. Inland taipans were not supposed to be black and white. Of more immediate interest to me than its unusual colouration was the flexibility of its mouth. I have held hundreds of other venomous snakes and always felt comfortable once I had them securely by the neck. This fella was different. Taipans have long heads so that their mouths can open wide enough to swallow large rats. Its roaming upper jaw was getting unnervingly close to my fingers so I quickly measured the snake: the snout–vent length was 1,165 millimetres, tail 187 millimetres, weight 452.3 grams and scale rows—twenty-three of course. With considerable relief I transferred the now agitated snake into a large glass-fronted display cabinet in my snake house.

When I dropped the first mouse into its cage, the inland taipan imperceptibly drew within striking distance and in an instant had struck at the mouse and then recoiled back to its original position. The only indication that the mouse had been bitten was that it immediately started dragging its legs and within seconds had fallen over. The taipan watched and then a minute or so later moved over and swallowed the dead mouse head first. Once or twice when provoked the taipan reared up in a double-S strike position, but it was easily the best-natured venomous snake that I had ever kept. Whereas some of the others bluff and huff and carry on it was almost as if the cool and confident taipan knew that it was revered by everything and everyone.

Within a few weeks of its arrival the taipan shed its skin and changed appearance. From jet black it turned to a dark bronze–brown with black flecks. The most noticeable change was that the belly took on an almost daffodil hue, while the head remained jet black. Such a pronounced colour change, particularly to a lighter colour as summer approaches, is also typical of some western browns in this region. The fact that its colour changed so dramatically highlighted the difficulties inherent in identifying inland taipans.

Another characteristic that was obvious to anyone viewing my collection of snakes was that the taipan was more active in cooler weather than the others. When the temperature dropped below about 20°C the western browns took on the alertness and mobility of a curled up string of sausages. The taipan was always alert but never stroppy regardless of the temperature. This observation tied in with its capture on a cold wet day and with the records from the channel country, which suggest that inland taipans only emerge from their cracks in the relatively cool days of autumn and spring. This may explain their apparent scarcity as inland taipans probably don't need to bask to raise their body temperature for activity, and can live out most of their lives concealed in the cracks.

Fortunately it was a cool February weekend that I chose to release the taipan back to its capture locality on the Moon Plain. The country was again in good nick with seeding Mitchell grass indicating that snake food would soon be on the increase. As it made its way off with its belly hardly touching the ground, the taipan occasionally lifted its head to check its surroundings. Several times the big snake meandered to within a metre of the release party without showing any signs of aggression or fierceness. Even when I changed its position for photographs it restrained from striking out or doing anything to warrant a name like the 'fierce snake'.

I watched it melt across the stony plain, nosing into holes in order to find a satisfactory home, its condition clearly enhanced by my parting gift of several house mice. I considered the irony that this was my only experience with an inland taipan in the wild, yet I felt that I now knew so much about them and the country they lived in. The big snake melted through the Dog Fence that was adorned with gold paper blown from the latest movie set. One beautiful, awesomely cool snake finally slipped into a pile of stream-strewn debris. This was the culmination of many weekends of driving, digging, searching and questioning. Somewhat in a trance after watching the snake for over an hour, I walked back to the car.

For the remainder of the weekend every twisted dropper or sinuous dark crack looked like another taipan. They appeared to be everywhere but I knew they weren't. As the Moon Plain dried up from its glory days in the mid-1990s, so the smattering of inland taipan sightings evaporated. Were they still out there or was the rash of taipans a brief passing phenomenon? I strongly suspect that like the plains rats they can hang on in imperceptibly low numbers. Now that the locals are aware of this snake my bet is that they will turn up a year or so after the next big rain. I certainly hope so anyway.

14
How long have you got?

An intense shot of pain snapped me out of my delirium. My heart raced as I sprung upright, sconing my head on a low branch in the process. Everything was going wrong. I felt vulnerable and scared, just like when I had seen those lights from the island in Lake Eyre South. Again I was hours away from the nearest people and although I had a car, my HF radio was not working. To make things worse I was also in pain.

A quick inspection of my nether regions revealed a juicy brown tick that was in the process of embedding its head into my flesh. Another sultana-like parasite was casually bumbling its way up my leg. Oh my goodness! The ticks freaked me out because I had never seen them in the desert before. I realised that the bush under which I had chosen to seek refuge was also the regular haunt of kangaroos, from which the ticks had probably originated.

Ten minutes earlier I had flopped down to contemplate my fate. It was the only shade that I could find. Elliot Price Conservation Park is not exactly the most densely forested tract of land in Australia. It is a windblown, largely barren peninsula that juts out into the vast expanse of Lake Eyre North. I had just commenced a fauna survey of this virtually unstudied and seldom-visited park. After chasing a large striped skink through, under and over several dry scratchy bushes, I noticed that my sock had become soggy from a stream of blood that was flowing down my leg. Although it did not hurt I became alarmed that I was losing so much blood from what appeared to be a little nick. Then I remembered the mulga snake.

Earlier that morning I had caught a mulga snake that seemed to occupy at least half of the Borefield Road. The seven-foot long snake was so strong that I could not hold its head with the conventional two fingers and thumb hold. Rather I had to clench it in a 'hammer hold' with my knuckles whitening under the strain. Although big snakes are generally easier to catch than smaller, faster ones, they are more difficult to control. The monster snake's tail was groping around for my legs to get a purchase. With its tail anchored the big fella would have been able to pull its head out of my grasp, which could have had serious consequences.

By the time I had fumbled around for a bag and wrestled its tail away from my legs, a considerable stream of venom had flowed from its mouth onto my hand and down my arm. Venom cannot be absorbed through the skin and I did not have any cuts on my arm so I was not at all concerned. The adrenalin-charged thrill of catching such a huge venomous snake swamped any worry cells in my brain. As I sped off up the road feeling like an invincible warrior I wolfed down the sandwiches that I had packed for lunch. An hour or so later I left the big snake in the care of Lyall Oldfield at his café at Marree, much to his chagrin as he is one of the outback's greatest serpentophobes. After another hour or so I had crossed the causeway over the Goyder Channel between Lake Eyre North and South, and was negotiating the sandy track to the start of the Conservation Park.

Mulga snake venom contains anticoagulant properties that have seen it used in preventing blood clots from forming during open-heart surgery. It dawned on me that I might have ingested a fair dose of venom when I ate my sandwiches with venom-soaked hands. I was not sure whether the venom could be absorbed through my gut wall or whether my digestive juices would break it down, but the flow of blood down my leg suggested that my blood was not clotting. If I had been envenomed internally I was in real trouble because the compression bandages that I carried for snakebites were of no use.

I decided that since I had not developed any other symptoms after catching the snake three hours earlier I would wash my hands and lie down in some shade. After scrabbling under the needlewood shrub I tried to convince myself that I was imagining the headache that was now pounding in my head and that it was concern, rather than snake venom, that had caused my increased heart rate. Then the tick bit me.

If I was going to succumb to snakebite I wanted to know that I would not be dined on by ticks so I dragged my swag out onto the dry salt lake. At least there wouldn't be any ticks or snakes out there. The burning sun was cooking my legs so I ended up getting inside the tarpaulin where I lay prostrate, sweating like a fat man in a sauna. From within my saturated swag I challenged myself with the same question that I had been asked countless times when discussing venomous snakebites: 'How long have you got?' Now the question was more immediate and personal. How long did I have to wait to know whether I would be dissolved from the inside out, and if this was to happen, how long would I survive?

When the amount of venom mulga snakes produce is considered along with their toxicity they are among the most dangerous in the world. However, the answer to the classic 'How long have you got?' question is not at all straightforward. The mice that were fed to the wild snakes that I kept for short-term study could provide only part of the answer. On some occasions a mouse appeared to drop dead before my snakes had even struck, as if they had been overcome by a toxic belch or a deadly stare. On other occasions the mice would hobble off and take over a minute to be subdued. The length of time it takes for a mouse to succumb to the venom depends on the amount of venom injected and where the mouse was bitten.

The situation with humans is even more complicated. Some bites are dry and no venom is injected at all. Dry bites are particularly prevalent when a snake is biting only for protection. I know of three occasions at Roxby when locals trying to catch snakes have been bitten, leaving clear fang scratches on their skin. None of these would-be snake catchers showed any symptoms at all. The bites were part of the bluff tactic that the cornered snakes were forced to employ to deter their 'predators'.

Just like with mice a bite directly into a human's vein obviously works faster than a shallow bite to an extremity. However, most Australian snakes have such short fangs that the venom usually travels under the skin in the lymph system rather than in the bloodstream. This is why a simple compression bandage like those used for sprained ankles can stop the flow of venom rather than resorting to the painful and dangerous tourniquets that were once recommended. By keeping the venom away from the vital organs with a bandage fashioned from a T-shirt, stocking or rag, the onset of any symptoms may be delayed for several hours. Unfortunately, bandaging would not buy me any more time if I had stupidly ingested the venom.

I was particularly annoyed with myself for having got into this predicament because I had often echoed the sentiments of other herpetologists that no-one in Australia should die of snakebite any more. Antivenom, which used to be called antivenene back in the days when tourniquets and magic potions were in vogue for snakebite treatment, has now been developed for all of our 'deadly' snakes. This antidote is extracted from horses that have gradually been exposed to the venom and neutralises the venom's effects. But assuming that some venom had reached my heart or brain, the 'how long' answer was still not straightforward. Since I was relatively large and fit I liked my chances of being more resilient than a child. But then again I am allergic to bees and had no idea if I was also allergic to snake venom.

Confused and afraid I considered writing a note in my diary in case I passed out. At the top of the note I would insist that no medical practitioner was to touch me without first ringing Julian White at the Adelaide Women's and Children's Hospital. Julian has a rare combination of medical training, toxinological research, a genuine interest in reptiles and a vast experience in dealing with snakebites. I have seen and heard of several incidences whereby otherwise competent medical staff who were naïve to contemporary snakebite treatment have made some potentially serious blunders. I am confident that Julian would know when to release the compression bandage, how to identify the snake and when to administer adrenalin and additional antivenom if required.

Ensconced within my sodden cocoon on the dry bed of Lake Eyre, I had one of those long hard looks at myself that my teachers urged me to undertake in my youth. What excuse could I muster for tempting fate by playing with venomous snakes? Why did I not just leave them alone? I owed an explanation to my family and friends if I did not pull through. In my head I composed a letter.

I can't help my fascination with snakes. They amaze and impress me. The way they move, the way they hunt, feed and mate are absolutely awesome. Any animal that can dig with their head, move rapidly on dry land without legs, catch fast-moving food without using arms and then swallow food twice the size of their head without choking, really flicks my switch. It's unreal that they can 'taste' the air to sense prey that they can neither see, hear nor smell.

Have you ever seen a snake melt across a road like an evaporating mirage? Have you marvelled at the fluid apparition that neither begins nor ends? Have you seen them suspend the front half of their body off the ground for minutes with more control and grace than a Romanian gymnast? Have you ever watched spellbound as they move imperceptibly towards their prey? Have you been dumbstruck by the explosive accuracy of their strike and release that leaves you wishing for a slow motion replay?

If you can't understand my infatuation with snakes it is probably because you have not witnessed snakes in the flesh, unaccompanied by the dramatic music and 'rattle-snake rattles' that are invariably associated with their TV appearances. I know that most people don't understand or appreciate their bizarre and fascinating features. That is the only excuse that I can make for the multitudes of morons who inevitably join a conversation with 'The only good snake is a dead one'. Yeah it rolls off the tongue well but it is one of the most naïve and puerile statements I have ever heard. It says far more about the orators than the subject. Do these same people think that the only good lion, shark or motorbike is a dead one too? Do they want to live in a sanitised, synthetic bubble with all fear and wonder, danger and stress, challenge and adversity removed? Leave me in the world of snakes any day.

A trickle of sweat seeped through my eyelashes and stung my eyeball. This soul-searching and justification had taken my mind off my current predicament. I threw off the tarpaulin and was relieved to find that my leg had stopped bleeding. Instantly my head stopped pounding and the world was a better place. I was not going to be tick or crow bait after all.

As I packed my swag and continued with my three-day survey I kept reflecting on the risks and benefits of studying snakes. Simply because we catch and study snakes other herpetologists and I assume the status of 'snake man', or less commonly, 'snake woman'. This places us firmly in the bizarre category of humanity, somewhere between a freak and a legend. Neither of these extremes are warranted. A snake has nothing to gain and everything to lose by approaching and attacking people or animals that are too big for them to eat. Invariably they will protect their delicate fangs and preserve their valuable venom for prey. Most snakes will avoid encounters with humans by slipping away before they have even been seen. Only when they are cornered will they typically 'attack' and even then it is usually an elaborate bluff designed to warn off an aggressor.

Strangely enough, snakes are far easier to catch and handle than many other animals that people handle regularly. Have you ever tried pulling a rainbow 'Rambo' lorikeet out of a mistnet? Their four needle-like claws and razor-sharp beak invariably draw blood while their demonic screeching severely stretches one's nerves and patience, yet parrot researchers are not revered for their bravado. Similarly vets are considered to be relatively mainstream, yet have you ever tried to pick up a reluctant cat or even an agitated rat? Lock me in a room with a taipan any day rather than with an aggressive pit bull terrier, breeding magpie or nest of inch ants. Even burly blokes who toss cattle on their sides for a living and dainty girls who perch precariously on racing or jumping horses have more dangerous jobs than mine!

Snakes are just as much a part of the unique Australian environment as koalas and platypuses, but no-one would suggest that these icons should not be researched. We know far less about snakes than we do about more 'conventional' study animals like mammals and birds. If it were not for the field research of comparatively few biologists like Australia's indomitable snake guru, Rick Shine, and the observations of hundreds of amateur herpetologists, we would understand even less of snakes' ecological role. We do know that since snakes are predators they may be more sensitive to changes in their environment than their more abundant prey. Snakes may also regulate the populations of some feral pests and provide us with a unique treasure chest of chemicals that doctors are only just beginning to uncover. Because some species are capable of inflicting fatal bites to humans our understanding of their lifestyles and habits has added importance.

Perhaps because few people are prepared to closely study snakes, there are so many questions that remain unanswered and this adds to their intrigue. It is incredible that we don't even know the identity of some of our largest venomous snakes. For example, there are at least two types of snakes that live around Roxby and throughout much of the outback that are called western browns. Yet these snakes are obviously, demonstrably different. In terms of their appearance, behaviour and preferred habitats they are as different as crows and magpies.

One of the local types of western brown grows to nearly two metres in length, is generally cardboard brown in colour and has variable patterning of black bands or spots down its body. In cool weather they darken to an almost blackish chocolate colour, while after shedding their skin in summer they may be almost white. These western browns never have a black head. The Arabunna people recognise them as being distinct and call these big snakes of the plains 'witjita'.

Two other forms of 'western brown' only reach about half the size of witjita and live almost exclusively on sandhills. One of these snakes is bright orange with a glossy black head; the other is a copper–green colour and never has a black head. As well as being much smaller than their gibber plains' cousins these dune snakes have small heads and lack the ability to flare their neck, cobra style. Whereas I have found up to four big hopping mice in the belly of a road-killed witjita the small sand dune western browns usually feed on smaller native Bolam's mice, house mice and lizards. Maybe they have remained small so they can fit down the holes of their prey on the dunes, whereas the much larger tableland snake can more easily find larger rodents and dunnarts in the 'crabholes' on the plains.

Roxby is not unique in hosting a confusion of brown snakes. Throughout Australia there are many recognised 'forms' of western brown snake. To add to the confusion other closely related and similar looking species including the dugite, peninsular brown, spotted brown and eastern brown sometimes coexist with this array of western browns. We still don't really understand how they differ from each other, nor do we fully understand their requirements, threats or potential value to humanity.

※ ※ ※

A consequence of our poor understanding of snakes is the number of furphies that have arisen about their behaviour. One such misapprehension is that sleepy lizards keep snakes away. Contrary to popular belief sleepies are not snake eaters; they are far too slow to chase snakes. Despite what you may have been told, sleepies do not latch onto snakes with their powerful jaws and refuse to release the crushed snake until they have their heads cut off. In reality the balance falls the other way. Large mulga snakes eat sleepies, even fully grown ones.

More believable myths have goannas or bearded dragons keeping snakes at bay. While I have seen both bearded dragons and goannas eating small snakes, one day I received a scary reminder that the reverse can also be true. I had been following the cotton thread that had fed from a spool taped to the tail of a bearded dragon. 'Spooling' animals is a useful method of retracing their movements and determining how far they roam, what they eat and where they shelter. The cotton thread that I was following had disappeared under a bush and had not re-appeared for a couple of days, so I got down on my hands and knees and peered into the small bush to locate the lizard.

Only a few centimetres from my nose a large mulga snake recoiled, sending me hurtling backwards in a shocked somersault. When I had recovered my composure and caught the snake I found a lump in its belly and the white cotton thread disappearing down its throat. Snakes, goannas and bearded dragons probably have a similar relationship to hyenas, lions and leopards on the African plains. A big one of either species will knock off a small one of any other. Although closely matched, large goannas probably hold the advantage because they are better able to eat a large snake than vice versa, and perhaps their tough skin protects them from the short-fanged snakes. Despite this apparent advantage, goanna–snake battles are rare and the presence of a goanna in the yard would have little influence on whether a snake may slide through or not.

Another common question, usually raised by a terrified householder who has had a snake retrieved from their yard, is whether having one snake is evidence that there may be a 'nest' or 'family' of snakes around. They ask, 'Will a big granddaddy snake avenge the capture of his offspring?', or 'Does the capture of a female mean that there will be a host of youngsters around?' The answer is invariably no.

Most snakes do not revisit their nest site or, in the case of livebearers, even remain with their young. However, sometimes snakes do aggregate at the other end of their breeding season.

One spring afternoon I received a panicky phone call from Jo Andryzak who said she had seen a snake in her backyard in Roxby Downs. When I arrived about ten minutes later I found not one, but two snakes that were 'joined at the hip'. Snakes cannot readily untangle themselves from their amorous encounters so I was faced with trying to catch a doubly aggressive snake with two blunt ends. What made it even more interesting was that the male snake was one of the small green types of western brown and the female was of the orange form.

Understandably for an animal that drags itself over the ground snakes are not endowed with external sex organs. While this makes life more comfortable for them it complicates a herpetologist's task of establishing their sex. Male snakes and other reptiles have two sexual appendages, one on either side. They can use one or the other of their hemipenes as they sidle up to their mate. The paired hemipenes are bizarre appendages that are typically adorned with spikes when engorged and everted. In the absence of love-handles or limbs these spiky hemipenes enable the male to hang on and explain why Jo's amorous snakes did not separate for a quick getaway.

The next day after the mating pair had separated and been released a few kilometres away, I received another call from Jo. She was in a terrible mess and about to pack her bags to leave town because she had found a third adult western brown in the same location. This snake turned out to be another male. What had probably happened was that the original female had left a trail of pheromones that indicated that she was receptive to mating. Both of the males had followed the trail of love potion possibly for hundreds of metres. Due to these pheromone lures it is more likely that several snakes will be found in the same area during the spring mating season than in summer when the juveniles disperse.

Establishing the identity and sex of the common snakes is difficult enough, but sorting out mythical or fanciful snake records from reliable observations is even more of a challenge. Perhaps the most bizarre story that has been told to me was by two senior Aboriginal men about the existence of a legendary and much feared whistling snake. This white snake, which is approximately ten-centimetres wide, vigorously defends its territory in hilly country. If an intruder does not heed two warning whistles the snake flies like a spear while whistling for a third time and inflicts a fatal bite.

Another Dreamtime snake that is slightly easier to comprehend is Yurgunangu, the red-bellied black snake that travelled through spring country from the Northern Territory down to the Roxby area. Red-bellied black snakes are familiar to many Australians because they live where many of us also like to live: in well-watered, cooler climates. These glossy black snakes with spectacular red flanks are not desert creatures at all. But pioneering pastoralist WBG Greenfield and Malcolm Mitchell from Muloorina Station have both seen several blackish snakes with red bellies near Lake Eyre. So too has Leo McCormack on Roxby Downs Station. I have not seen these crimson-bellied black snakes that tenuously maintain the links with Yurgunangu. I can only assume that they must be unusually boldly marked western browns or inland taipans.

Neil McLean, ex-head stockman of Anna Creek Station, told my historian mate Phil Gee that he once saw a death adder near George Creek, just north of William Creek. This would most likely be the desert death adder rather than the more common death adder that lives around the coast. Jeff Boland also reckons he has seen a death adder squashed on the road just south of Coober Pedy. Judging by his success with locating inland taipans Jeff's sighting has to be taken seriously, even though there have been no verified records of desert death adders from South Australia for the past eighty years.

Stewart Nunn, also from Anna Creek, along with a couple of ringers claim to have seen large docile snakes resembling carpet pythons in the sand dunes near William Creek. Carpet snakes have not been officially recorded from within cooee of William Creek, nor from similar dune country without plenty of hollow trees. Stewart has also occasionally seen woma pythons in these same dunes and in 2001 a stockman at Anna Creek called Butsy could not avoid a huge woma that took up much of the road. Old Max Hall who lived on Roxby Downs Station from 1931 to 1940 also remembers seeing womas in the district. The blackfellas ate three or four womas while Max was on the station. The country around Sisters Well, which is only about thirty kilometres from town, was a particularly good place to find womas or their distinctive tracks. Max remembers them being harmless snakes but was not too impressed with their taste. A couple of tantalising reports are the only suggestion that these huge pythons still live around Roxby.

My best chance of finding a Yurgunangu, woma, carpet snake, or desert death adder presented itself when WMC dug a pipeline trench all the way from Lake Eyre to Roxby. Up to ten kilometres of this two-metre deep trench was open ahead of the pipe layers at any one time. Along with a rotation of eager snake catchers I patrolled the length of this trench every morning and some nights before the not-so-eager construction crew started work. Although ramps were constructed every few hundred metres for animals to escape, the snakes seemed to prefer to stay in the trench. The snakes would meander down the trench feasting on mice, lizards or other snakes that had also been trapped.

Although we pulled nearly 600 snakes from eight different species out of the trench we unfortunately did not encounter any death adders or large pythons. However, as if to prove what a great snake trap it was, we did find a couple of rarely seen Stimson's pythons and many snake-like Burton's legless lizards. Legless reptiles were not the only interesting finds in the trench. Among the nearly 3,000 individuals and forty species of lizards we recorded several desert skinks that had not been recorded in the region. In a section of the trench that passed through taipan country we found both gibber dragons and a new population of plains rats with unusually white tails. Desert mice, planigales and even trapped cows boosted the mammal inventory to over a dozen species from this unique 150-kilometre trap.

Despite these exciting discoveries the real stars of the trench were the curl snakes. Their name derives from their tendency to coil when provoked and then to lunge out, throwing their full body forward at their aggressor. Curl snakes were the mini-vacuum cleaners of the trench and it was rare to collect one from which a partially digested dragon, gecko or smaller curl snake could not be carefully enticed from its swollen gut. These little grey snakes with a pearly white belly and dark flattened head have the most disarming ability to move their eyes independently. Like inland taipans, curl snakes are still active in cool weather and we retrieved hundreds of these nocturnal snakes in June while our breath misted in the night air.

The pipeline trench revealed just how abundant snakes are in the outback, which in turn highlighted how well they normally avoid contact with humans. Every summer hundreds of inquisitive children come within metres of unaggressive snakes at Roxby, while many of their naïve parents rave about the danger of supposedly sinister serpents. While a handful of these children will hopefully retain their wonder and compassion for our wildlife, others are destined to acquire the irrational fear of snakes from their ill-informed parents. Unfortunately, as fewer and fewer people in today's sanitised society encounter snakes frequently, the pathetic catchcry of 'good snakes being dead ones' is likely to gain momentum, rather than being banished to history where it belongs.

15

The climatic tightrope

It is often the dream of many biologists to discover new species or to rediscover ones presumed extinct. Finding new species of vertebrates is a rare occurrence these days unlike the 'glory' days of the pioneering taxonomists like Linnaeus and Gould, when every second animal or plant they encountered was looking for a name. Most new species of Australian vertebrates recently described have been cryptic critters that were previously grouped with similar-looking types. These 'new' species may be distinguished by factors such as their genetic make up, their call or in the case of one bat—the size of its penis. Many are only recognised as being distinct following detailed analyses in museums months or years after their collection. Therefore when young Steve Delean picked up a distinctively different type of knobtail gecko just south of Pimba in 1981, it was cause for much excitement in the herpetology world.

Steve had been on a field trip with four members of the South Australian Herpetology Group. They were searching for knobtails, which are large geckos with a characteristic knob on the end of their tails. On the final night of their trip to the region north of Ceduna where they had recorded three species of knobtails, they went spotlighting along a dirt road on the eastern side of the Stuart Highway near Pernatty Lagoon. The knobtail that they found was an adult female. Ironically it was caught where a juvenile gecko had been collected a decade earlier: this gecko had erroneously been catalogued as a midline knobtail (*Nephrurus vertebralis*). The young herpetologists had seen midline knobtails days earlier and they instantly recognised that the gecko that Steve had found was different.

Under the guidance of Chris Harvey, president of the Herpetology Group, Steve and his mates subsequently mounted several expeditions to the area. In the sand dunes around Mt Gunson and the northern tip of Pernatty Lagoon they collected several more of these big spunky geckos. Chris subsequently described the Pernatty knobtail and gave it the scientific name of *Nephrurus deleani*, in recognition of Steve Delean's efforts. In 1983 when the Pernatty knobtail was formally described, Chris wrote 'specimens have been collected only from the acacia vegetated sandhills immediately north and west of Pernatty Lagoon, despite extensive searching to the north and south'. No new information had been published explaining the limited distribution of the Pernatty knobtail, when I started trying to understand their story in the early 1990s.

Because of its restricted range the Pernatty knobtail is considered to be vulnerable to extinction. It is one of only four geckos and fifty-two reptiles in Australia that has been recommended for this status by the Action Plan for Australian Reptiles. Along with the bronzeback legless lizard, Pernatty knobtails were the only vulnerable reptiles found within hundreds of kilometres of Roxby. I wanted to establish the extent of their range and why they were apparently restricted to such a small patch of sandhills. My approach was to start looking where they had already been caught, familiarise myself with their preferred habitats and activity periods and then push back their known distribution limits until I located the barrier to their dispersal. The first night I looked for them on the tall dune on the southern shore of Lake Windabout near where Steve caught his Pernatty knobtail. I enjoyed immediate success. For such a supposedly rare animal I was amazed that I found them so easily.

Pernatty knobtails are a honey–orange colour with indistinct pale and dark stripes and spots. As is typical of knobtails, when approached they stand up high on their legs arching their back to maximise their bluff and squeak, like a toy puppy. Like the other members of its genus the Pernatty knobtail possesses the endearing large eyes and charismatic appeal that sets its apart from those species that don't make it into magazines. I was hooked and I spent many nights wandering around the dunes south of Pimba searching for these geckos.

I trapped and spotlighted in several different habitats between Pernatty Lagoon and the Stuart Highway. Pernatty knobtails lived in both the pale yellow sand by the shore of the lakes and the darker orange sands further away. Rather than being restricted to dunes with acacias they were also found in areas vegetated by canegrass and hopbush. I could not find anything unique about their habitat that would restrict them from living in dunes or sandplains in a wider area, so I started searching further out.

From many years of trapping I knew that Pernatties were not found around Roxby, 100 kilometres to the north, so I looked in the dunes halfway back towards their known range. It took me several attempts to find any knobtails on the dunes around the Arcoona Homestead and Giles' Hut, to the north of Pernatty Lagoon. When I eventually found a couple of them scurrying down rabbit holes they proved to be the three-lined knobtails that were common in the dunes further north. The most noticeable difference between the Pernatty knobtail and the more common three-lined knobtail (*Nephrurus levis*) is that the tail of these new geckos was much more slender. By contrast, the swollen fat storage tail of the three-lined knobtail can be nearly as wide as its head. When I looked at the dunes inside the newly formed Arid Lands Botanic Gardens at Port Augusta I caught a couple of three-lineds but no Pernatties. This site was just over 100 kilometres south of their known range so it appeared that they really were boxed in.

I decided to try the dunes on Island Lagoon that were visible to the west from the Stuart Highway. A handful of Pernatty knobtails in the canegrass dunes was the best twenty-seventh birthday present I could hope for. The finds were significant because these restricted geckos had now been found adjacent to another lake. Island Lagoon has pockets of dunes at both its northern and southern extremities, and since I had just caught Pernatty knobtails on the centre west I expected to further increase their known range.

The next night I searched the dunes adjacent to where Eucolo Creek enters the northern tip of Island Lagoon. Although the country looked identical to those places where I had been successful a mere thirty kilometres to the south, I only found three-lined knobtails. A pattern was starting to emerge. It seemed that sand dunes supported either Pernatty knobtails or their fatter-tailed relatives, but not both. Unfortunately I had no idea why certain dunes were occupied by one species while adjacent, virtually identical dunes were occupied by the other.

One day I was pondering this conundrum while flying down from Roxby to Adelaide. The flight path of the little flying cigar that I was crammed into went straight over the Pernatty knobtails' dunes, so I craned my neck backwards and forwards to check out suitable localities for 'ground truthing'. One particular group of dunes screamed out to be checked. They were isolated on the shore of Island Lagoon almost halfway between the locations where I had recorded each kind of knobtail. On both sides of these dunes were kilometres of hostile gibber plains or salt lakes. Could Pernatty or three-lined knobtails colonise these isolated dunes across this inhospitable habitat? The greatest impediment to finding out which knobtail lived there was that the dunes were next to a couple of gigantic white balls, which I knew meant trouble.

The next weekend I drove to the Area Administrator's office at Woomera as arranged. Bernie McCarthy, who took his job very seriously, informed me and my accomplices that there was no way that we would be able to wander around the dunes so close to the Americans' Nurrungar installation. The Yanks had obviously convinced at least one Australian that this 'Big Brother' spy base was of utmost strategic importance, so much so that a carload of gecko catchers within ten kilometres would unduly compromise world peace into the next century. Accompanying me that day was Zoë Bowen, an environmental officer at Roxby who would never consider even parking in a No Standing Zone, and Mick Evans with his three feral boys. None of us were any threat to the US Special Forces. Eventually Bernie agreed that he would escort us to the base of Vanguard Hill, a safe distance from Nurrungar. We could camp there and look for geckos, but were not allowed to follow the dunes north towards the huge white geodesic domes.

'Yeah, righto Bernie … whatever you say,' were our parting words.

Bernie left before dark and so did we, heading north. Within an hour we had caught nine Pernatty knobtails. We had again increased their known range and, more importantly, had proven that they were able to colonise sand dunes isolated by stony tableland.

Together with a rotation of different mates I searched successfully to the east and found Pernatties on the northern and southern tips and in some dunes on the eastern side of the shallow lagoon from which they take their name. We also found them to the south on the edge of Ironstone Lagoon and on the western shores of Lake McFarlane. The latter locality gave me perhaps the biggest buzz because it was about sixty kilometres from their 'original' locality and represented another relatively isolated patch of dunes. This growing list of Pernatty knobtail localities was more than just a satisfying tally; each separate population that we discovered provided considerable additional security to the survival of the species.

For a while it seemed that we would find Pernatty knobtails on any dune we looked at but the next weekend we ran out of luck on Mahanew Station. I also ran out of luck in the dunes near Woomera. Even though we had greatly extended their known range, nearly the entire reach of this species could be seen on a clear day from the summit of Mt Gunson. They remained restricted in distribution and as such were clearly a vulnerable species. The key to their long-term conservation could rest with an understanding of why they had not colonised other nearby dunes.

※ ※ ※

As long as they had sand to burrow into, Pernatty knobtails were obviously not fussy about their habitat. Neither were they affected markedly by the condition of the land, being found in areas severely degraded by both sheep and rabbits. I could not rationalise their restricted distribution with an inability to disperse across tableland country, as they could clearly occasionally breach these stony plains. I was not aware of any predators that would target Pernatty knobtails outside their present range and by looking at their turds we found that like other knobtails they ate a range of food including scorpions, spiders, insects and other geckos. So it was also unlikely that they were limited by food resources. In fact the most consistent finding was that they did not seem to coexist with three-lined knobtails, their fatter-tailed cousin. Maybe there was something more than coincidence in that?

In some wetter parts of arid Australia three-lined knobtails coexist with another slender-tailed species called the smooth knobtail (*Nephrurus laevissimus*). In such areas I have caught the slender-tailed gecko mainly on the sand dunes and the thicker-tailed three-lined knobtails on the interdunal flats. Yet when they were the sole knobtail in a region both the three-lined knobtail and the slender-tailed forms occupied a wide range of sand habitats. Apparently these two species either avoided each other or discouraged each other from their distinct habitats. Ecologists refer to this antagonistic relationship between two similar species as competitive parapatry.

Many pairs of species that have parapatric, or non-overlapping, distributions cannot coexist because they have nearly identical food, shelter and climatic requirements. The best-adapted species for each patch of dunes would in time displace the less-suited species. The boundary between the two species was in most places simple to determine as they were found in distinct sand islands. Later on, a research student Harald Ehmann found a zone where both species coexisted along the boundary of their distributions. This could well be the coalface of the tug of war between the two species.

The question that haunted me, whether sitting around a campfire on a sandhill or flying over Pernatty Lagoon, was why Pernatty knobtails were apparently favoured in one location while their cousins dominated in adjacent, virtually identical sandy areas. The most obvious clue was that the answer had something to do with their tail size. The skinny-tailed Pernatty was sandwiched to the north and south by the fat-tailed species.

When the distributions of skinny and fat-tailed knobtails are compared it becomes apparent that the slender-tailed forms live in the wetter areas of the arid zone, whereas old fatty three-lined knobtails can live in the driest dunes of the Simpson and Tirari deserts. As I mentioned before, when they occur together the slender-tailed knobtails tend to monopolise the sand dunes that generally support the most food. Sand dunes also tend to be warmer than the interdunal flats. Just as dunnarts tend to have fatter tails in harsher environments where they often need to maintain energy stores for lean periods, the fat tail of the three-lined knobtail is probably an adaptation to unpredictable and tougher arid environments. On the other hand, a slender tail may be a more convincing lure for prey like spiders, scorpions and smaller geckos. Therefore, Pernatty knobtails may be able to out-compete three-lined knobtails where food is more abundant or weather conditions more suitable. This theory has Pernatty knobtails winning the competitive battles when the larders are well stocked and three-lined knobtails being better able to persist with barer pantries.

It is rainfall that primarily determines how well stocked a knobtail's pantry is. Vegetation growth and insect abundance is generally correlated with rainfall in arid regions. Rainfall declines and is more erratic to the north of Pernatty Lagoon, which may explain why the three-lined knobtails out-compete Pernatty knobtails there. The eastern distribution of Pernatty knobtails is apparently limited by Lake Torrens, Australia's longest lake. To the west the habitat changes and Pernatty knobtails may be restricted through competition with other knobtails. However, their southern limit is possibly another climate-related barrier.

Day time temperatures generally increase from south to north in Australia due to direct latitudinal factors and because the southern and coastal areas are more likely to be cooled by cold fronts than northern inland localities. A gecko with relatively small fat reserves needs to feed regularly and is therefore disadvantaged to the south where temperatures are often too cold for gecko activity. A three-lined knobtail that can sit out more cold days by relying on the store of lipids in its tail, may therefore be competitively advantaged around Port Augusta to the south.

If indeed the Pernatty knobtail is balancing on a climatic tightrope strung between the cold south and the dry north, what implications does this have for the continued survival of one of Australia's most restricted lizards? The climate of our deserts is predicted to change as a result of global warming. Dunes that are currently optimal for one species of knobtail may no longer be suitable following a shift in temperature and rainfall. No-one can predict exactly how the delicate balance between Pernatty knobtails and their three-lined cousins will respond to climate change. Steve Delean's descendants are unlikely to find this unique gecko in similar numbers and at the same localities that he did. What's more, knobtails are not the only local lizards that are likely to change their distribution in the near future.

※ ※ ※

Rock dragons are medium sized, often brightly coloured lizards that inhabit rocky outcrops. From the Mount Lofty Ranges to the Northern Territory border, South Australia hosts at least six species—

twice as many as any other state. I say at least six species because one rock dragon, *Ctenophorus tjantjalca*, was only described from the far north in 1992 and I am quietly confident that another population south-east of Coober Pedy may yet prove to be another species.

Arguably one of the most brilliantly coloured of the rock dragons is the male of the northern form of the peninsula dragon. These dark lizards with bold red bands on their flanks live around the perimeter of the Arcoona tableland and hence are known locally as Arcoona dragons. Although they are a separate species, Arcoona dragons closely resemble the red-barred dragon that lives in the northern Flinders Ranges and that extends to within about eighty kilometres of the Arcoona dragon's range. Like their close relatives, Arcoona dragons are athletic lizards with hind legs that a high-jumper would kill for. I have watched in awe as they jump thirty centimetres vertically and half a metre horizontally to capture flies and bees attracted to blue rod flowers.

What I find most interesting about Arcoona dragons is not their appearance, their athleticism or where they live, but where they don't live. Rock dragons are generally not too fussy about their habitat. Dominant males usually command the best rocks for displaying and surveying their territory. These prime locations are typically adjacent to crevasses where they can seek shelter if alarmed. Others occupy less ideal habitats and some even live on isolated boulders. Many of the rocky creeks that flow off the Arcoona Tableland support populations of Arcoona dragons. However the rocky creeks and even the cliffs in the centre of the tableland are generally devoid of the lizards.

While I was wandering around these lakes counting ducks I always kept my eyes open for Arcoona dragons. The cliffs supported other rock-dwelling reptiles such as the bizarre spiny-tailed or Stokes skink, wall skinks and Stimson's pythons, and I was convinced that the habitat was suitable for rock dragons. I have caught Arcoona dragons a reasonable distance from large rock outcrops and could see no reason why they would not have been able to colonise these lakeside cliffs, especially since other less mobile rock-dwellers had made it. This dilemma remained a mystery to me for several years and the apparent answer was most surprising.

Roy Ebdon and I had paddled out to Coorlay Lagoon to band some chicks from a colony of 300 gull-billed terns. While we were moving through the colony I noticed that many of the nesting scrapes had been constructed of tern turds and that many of these turds contained undigested lizard tails. Although grasshoppers appeared to be their main food I found the remains of thirteen juvenile bearded dragons. Gull-billed terns only visit northern South Australia when the lakes fill and I had wrongly suspected that like other terns that they fed on aquatic invertebrates, small fish and maybe dragonflies.

I started paying more attention to these black-beaked white birds and noticed that they typically flew low over the shrubland around the lakes searching for grasshoppers and dragons. The dragons that the terns had been feeding on at Coorlay bask on the tops of bushes and hence are easy pickings. This may be an interesting observation but what relevance does it have to Arcoona dragons or climate change?

Because they bask on exposed rocks Arcoona dragons would presumably be snapped up by gull-billed terns. Predation by terns would be particularly intense since the rock dragons inhabit the cliffs immediately surrounding the lakes. Roy and I recorded gull-billed terns nesting on several different lakes. By contrast, terns were rarely seen on the large salt lakes on the margins of the Arcoona tableland, which dry more rapidly and do not have as many suitable nesting islands. My theory is that Arcoona dragons are not found around many of the Arcoona lakes because whenever the lakes fill terns wipe out any populations that have bridged the gap from other rocky habitats. Bearded dragon populations are less vulnerable because the adults are too large to be taken by terns and hence local hatchlings can replace the young dragons that are taken by terns in wet years.

To prove that terns restricted the distribution of the dragons, the fate of several populations of Arcoona dragons would have to be documented following the arrival of a tern colony. This could take years or decades of close scrutiny. In the same way, proving how pelicans locate inland fish resources and why Pernatty knobtails are restricted to a limited patch of sandhills would take more time and resources than I could justify.

Charles Darwin is probably the most celebrated lateral-thinking field biologist of all time and once said, 'I am a firm believer that without speculation there is no good and original observation'. To take Darwin's stance further, why not stick your neck out and propose untested speculations so that someone else may be motivated to provide supporting or conflicting evidence?

You can take or leave my theory on gull-billed terns restricting the distribution of Arcoona dragons, but concerns about the effects of climate change on restricted animals and plants are almost bullet-proof. Terns are not the only predators that visit the arid zone after rain. Silver gulls, which should more correctly be called 'scab' gulls, also flock inland whenever the lakes fill. Scab gulls are highly efficient predators, which may shock observers who have only seen them dining on chips in McCarparks or occupying prime-viewing locations at the cricket. They are believed to be the most significant predator of another species of dragon that is restricted to arid zone lakes—the Lake Eyre dragon. When their salt lake habitats flood these dragons that normally live on the dry lakebeds are forced to take refuge on the shore. At this time they are readily picked off by gulls, terns and other birds.

A few decades ago this did not happen. Populations of scab gulls have boomed since our wasteful society has provided them with a regular and exorbitant food supply. Most outback towns now provide the key attributes of a scab gull paradise: a takeaway store, dump, sewer pond and cricket oval. Large colonies of scab gulls now have ready access to inland lakes and they invariably flock to islands adjacent to nesting colonies of other birds, often with disastrous results.

Clive Minton, an enigmatic Victorian wading-bird fanatic, has followed the recent breeding success of banded stilts throughout Australia. Banded stilts are unusual in that unlike other waders their breeding seems to be reserved for rare occasions when most of the species congregate on individual salt lakes. In the same way that some of our desert trees are under pressure because they only recruit occasionally, the apparent high densities of banded stilts that appear intermittently at salt lakes may conceal a bleaker population trend. Although not desperately uncommon, Clive and his team of waderphiles have documented alarming evidence that suggests that the survival of these unique flamingo-like birds is perilous.

Of the three main nesting episodes that have been recorded over the last decade, nearly all of the banded stilt chicks have been taken by scab gulls before they had a chance to leave the colony. In his booming Pommy accent Clive recalls how he even witnessed gulls pulling and bullying adult stilts from their nests on Lake Eyre in order to steal newly hatched chicks. The stilts have no chance when colonies of several thousand gulls nest nearby. With little or no recruitment the population of banded stilts, a species restricted to Australia, is getting older and older and will eventually collapse from over 100,000 individuals only a decade ago. It can happen, because, like the ill-fated passenger pigeon, the banded stilts rely on their dense colonies for finding food and protection.

Fortunately, Clive's work stimulated a successful silver gull culling program on Lake Eyre in 2000 that enabled the banded stilts to enjoy a second far more successful breeding. No doubt this culling also gave the Lake Eyre dragons some respite from the onslaught, although because they are reptiles and not birds, the lizards attract less interest and monitoring. Controlling one native species to protect another always involves some soul searching, but in the case of the stilts this work was absolutely necessary and by no means unprecedented. Burgeoning populations of herring gulls and black-backed gulls are now poisoned in several European countries because they have been responsible for decimating the eggs and nestlings of ducks and other birds. Like our silver gulls, these gulls have also increased in number by several orders of magnitude in the past century.

What has tern counting and gull control got to do with climate change? Lots. The level of carbon dioxide in the atmosphere is now a staggering twelve times greater than it was when scientists first realised the significance of global warming. Principally through the burning of fossil fuels, modern society is contributing to increasing concentrations of carbon dioxide and other 'greenhouse gasses' in the atmosphere with major ramifications for the Earth's climate.

Meteorologists expect that global warming will cause the climate of central Australia to become warmer. Ecologists are already concerned about the effects of increased temperatures. At least half of the red kangaroos around Roxby died along with many other native animals during a heatwave in January 2001, when the maximum temperature exceeded 45°C for seven consecutive days. If these heatwaves become more frequent or severe it is likely that several of our outback icons will no longer be able to survive in this region. Fewer cold snaps could also undo the lid on plagues of locusts and flies with further flow-on effects to many plants and animals.

A less apparent implication of global warming is the predicted increase in flooding rains. More floods mean that the lakes will contain water for longer and the gulls and terns will become more abundant and spend longer in the arid zone to the detriment of the dragons, stilts and other arid zone species. Although gull control assisted the banded stilts at Lake Eyre there is no way that we can provide management solutions for all of the ecological shake-ups that may result from climate change. The weeds, pests, competition and predation from invading native species could well leave the outback in a very different state to what we see today.

PART FOUR
If critters could talk

16

27 km NE of Lyndhurst

I was standing as quiet as a tree stump. The fingers of my left hand were brushing over my eyes. My right hand held my binoculars and was also swishing flies from my ears. I knew it was here somewhere probably zigzagging away behind bushes to escape. If I made any sudden moves at the flies the subject of my attention would surely flee, but because of the swarming sweat suckers I could scarcely hear its feeble peeping. A poverty bush prickle had worked its way into the sole of my right foot forcing me to limp. The prickle had got in because I wasn't wearing socks. Being sockless is one of the few ways I can distinguish work from play. Socks are an expected part of work attire but are a nuisance when slipping my boots on and off to walk in water or boggy mud.

Eventually I saw the bird that I was after and checked that it had the cinnamon colouration and short orange tail that distinguished it from the chestnut quail-thrush. At last I could have a decent swipe at the flies and shake my boot off to remove the prickle. Cinnamon quail-thrush are common birds of the gibber plains and I had seen them thousands of times. But this record was better than most. As soon as I had scribbled CQT on my hand, I turned my attention to a patch of mulgas where I wanted to add a SHE to the sweat-smeared tally. Singing honeyeaters were even more common than cinnamon quail-thrush. I had seen literally millions of 'singers', and they were not even a particularly attractive bird. They are a drab olive brown with a flash of yellow and white on their face. Even their name is misleading because their rather harsh cheeping does not constitute much of a song. The honeyeater I was chasing had not been banded and was not integral to any of the ecological studies that I was conducting. I was getting frustrated, prickled and harassed by flies on a gibber plain north of Oodnadatta for fun.

I was having so much fun because I was twitching. Many scientists consider that 'twitching', or marking off a list new bird species sighted, is the lowest form of ornithological endeavour. So I had also appealed to serial twitchers to collect information on the seasonal trends, diet, behaviour and breeding of the birds that they were so determined to record. 'Try to work out why kestrels rarely hover in the arid north like they do in the agricultural regions, or why pardalotes call in the middle of the day when the other birds are quiet', I proposed. Twitching is challenging but answering questions could be far more rewarding.

After camping at the Carpamoongana waterhole on Hamilton Creek for a couple of days, I filled out the pink Birds Australia Atlas Record Form for the site while sitting on an esky in the shade of a red gum. These forms are completed by thousands of volunteers from all over Australia so that the distribution and possibly the abundance of birds can be assessed and monitored. Instinctively I tallied my list of species. Fifty-three was not bad but there were several common species that I had not recorded. The little circles next to Richard's pipit, singing honeyeater, black-faced woodswallow and cinnamon quail-thrush had not been checked off. Less than 200 metres from where I had been sitting I could look over a stretch of gibber plain to a patch of mulga trees, and I knew that I should be able to get all these species within a short walk. I had been lured into the self-gratifying trap of twitching a long species list which was why, having already scrawled BFWS, PIP and now CQT on my hand, I was now battling with the flies and prickles to find a SHE.

The previous day, along with Katherine, budding ecologist Kelli Jo

Lamb and Whitey, another mate from Roxby, I had completed several twenty-minute bird inventories in patches of representative habitat on Lambina and Hamilton stations. When compared with the widespread inventories compiled for the initial *Bird Atlas* survey between 1977 and 1981, these short, standardised bird inventories were of limited value in comparing changes in bird distributions. But Geoff Barrett, who was heading up the current *Bird Atlas*, was committed to the short surveys because they provided a repeatable snapshot of the relative frequency that birds were recorded in a range of different regions, habitats and land uses. For the first time, the Birds Australia scientists would have a baseline to track how the status of birds changes at a finer, more precise scale than the previous wide-scale survey.

I took on the role of the coordinator for the *Bird Atlas* counts for northern South Australia in 1999 with some trepidation. I was concerned about the accuracy of the *Atlas* due to the number of volunteers that were involved, many of whom were unfamiliar with local birds. I was also aware that several respected ornithologists were not involved because they did not have time to fill out the sheets, or were concerned that other scientists may 'poach' their information. Could such a huge ad hoc survey really provide useful information?

I had used the original *Atlas* repeatedly to check on the significance of distribution and breeding records of birds that I encountered. If for no other reason than to stimulate the interest of birdwatchers and to encourage them to observe more thoroughly and carefully, the *Atlas* is a success. But the *Atlas* is unable to provide all of the information to which its coordinators aspire. Contributors were urged to compile their bird lists representatively in the variety of habitats that they encountered. Yet as I was vetting the records that came in from bird watchers across the country, a handful of localities featured on just about everyone's outback odyssey. The Mungerannie Boredrain, halfway up the Birdsville Track, and the Cullyamurra waterhole near Innamincka were two such magnets for many outback bird atlassers. It is difficult to expect birdos to conduct surveys on open gibber country when there are a couple of brolgas standing in the boredrain just up the track. Similarly, why camp on a barren sandy flat when you can roll out your swag under mighty red gums next to a permanent waterhole? As a result dotterels, crakes, ducks and treecreepers were recorded from the far north-east of the State nearly as often as resident stony desert birds such as fieldwrens and brown falcons, which are far more abundant in the region.

Boredrains and waterholes were not the only 'birdo magnets' that introduced potential bias into the *Atlas* survey. Perhaps the most obscure but commonly surveyed site was located '27 km NE of Lyndhurst'. That was the only name that *Atlas* observers gave this spot. There was no enticing waterhole yet dozens of birdwatchers stopped on the side of the Strzelecki Track to compile bird lists. The inventory from this site was never exceedingly long, and included a group of birds that were fairly typical of shrublands on hard soil; white-winged fairy-wrens, cinnamon quail-thrush, black-faced woodswallows and nankeen kestrels are hardly the type of birds that would lure bird enthusiasts from around the country. Another thing distinguishing the pink sheets that came in from this site was that rarely was the preferred two-hectare, twenty-minute count conducted. Atlassers were given the option of surveying larger areas for greater lengths of time and birdos invariably spent an hour or so at 27 km NE of Lyndhurst, and they ticked the five-kilometre radius search area box. They were clearly coming to this location and wandering around looking for something special.

That 'something special' was a little brown bird about the size of a zebra finch. This unobtrusive, rather bland-looking bird would not even divert the eyes of the ornithologically challenged, like perhaps a gaudy parrot or elegant brolga would. A novice birdo would be annoyed by the little bird that was frustratingly quiet and flighty, taking off quickly over the low rolling plains before any distinguishing features could be seen. The beginner might return to their car and announce that along with the confusingly named black-faced woodswallow, which sports a mask while the masked woodswallow has a full black face, that they had also seen a pipit and a wren and been happy with their walk through the bluebush. Not so the people that were sending me pink forms. It was the tantalising glimpses of the little brown bird that they were after. That is why they drove from Victoria, New South Wales and even Western Australia to that special place, half an hour north of Lyndhurst's 'Elsewhere' Hotel.

Ten years earlier I had been there myself. I had taken a few days off to help Lynn Pedler with his latest project. More to the truth, I was travelling up the Strzelecki Track to learn from a bloke who had a better feel for the local birds than most. The first time I met this quietly spoken farmer from Koolunga, a couple of hours north of Adelaide, I was astounded at his rapport with birds. I had been studying some rare firetail finches at the Coorong and had collected a lot of information on their diet and habitat, but had never managed to get a decent photo of them. While I had chased these elusive birds around with my camera, eventually giving up when they flushed away, Lynn strolled in with a huge lens and a far superior approach. He looked around for a suitable shrub with the sun in the right spot, a suitable dark background and a low tree where he could stand under cover. He then proceeded to whistle.

Being able to recognise and imitate bird calls are probably the most difficult and useful skills that an ornithologist can possess. Lynn was a master. He called in robins and honeyeaters and finches that dutifully flew to the perch that he had selected because they were eager to determine who had invaded their territory. For a while I sat and watched from a distance spellbound by this display. Then I snuck off into the scrub and tried to pucker my lips and twist my face around bird squeaks. I felt like a novice trumpet player trying to imitate Louis Armstrong.

Lynn had spent a lot of time chasing arid zone birds and was the perfect bloke to teach me new tricks. After I had driven the prescribed distance up the Strzelecki Track from Lyndhurst I joined Lynn in walking and carrying mist-nets and poles; our ears peeled and binoculars ready. Not only were we planning to observe one of Australia's rarest birds, we were also attempting to catch some as well. All afternoon we walked up and down the hills without luck, despite being in the best place in the world to see them.

Soon after we headed off the following morning Lynn stopped abruptly. He had obviously heard a bird and was staring ahead, so was I. After a while I saw movement through my binoculars. Two cinnamon quail-thrush flew off low to the left. They always fly low, as if they are afraid of heights. I looked over at Lynn, expecting him to continue walking after the false alarm, but he had not. He was still looking straight ahead and counting out loud ... two, three, four. Clearly I had missed something so I raised my binoculars again.

Next to where the quail-thrush had been flushed was a small flock of fawn-coloured birds hopping around on the ground. One that was facing me had a whitish breast and my mouth dropped open when I saw the distinct orange–brown band. Jackpot. I was looking at my first chestnut-breasted whiteface. As the flock nervously moved off in the direction of the quail-thrush I listened to their plaintive trilling call.

My enchantment with these birds was broken when Lynn grabbed the poles from my hand. He asked me to scoot around the departing flock and muster them back towards him. I raced out wide and came back to a spot where I thought I would have the birds between me and the net, only I couldn't find them again. I turned to look where I heard the distinctive peeee of a quail-thrush and immediately saw the little flock of whitefaces. I was learning an important lesson about finding chestnut-breasted whitefaces. First find the larger, louder quail-thrush and then start looking hard. As I flushed them back towards Lynn who was feverishly putting up the net I counted seven birds.

One peeled off from the group before I got them back to Lynn and four of the remaining six saw the net and dived under it. The fine ten-metre long net supported by a pole at both ends was billowing in the wind and clearly visible. To our relief two birds panicked and were caught. Only in the hand was the subtle beauty of their plumage apparent. The distinctive chestnut-coloured band of one of the birds was edged with black, whereas the other's band merged seamlessly with its white breast. They were the sixth and seventh chestnut-breasted whitefaces that had ever been banded. Not only did our prize captives receive a uniquely numbered bird band, they were also fitted out with three coloured plastic bands so that they could be recognised again without recapturing.

About an hour later we again heard the distinctive twitter and saw three more birds, two of which were colour-banded from a previous trip. Even later that afternoon we located another flock of ten white-faces, of which we managed to net a world record of four, as well as a thick-billed grasswren. I remember thinking that there were not too many bird banders who could boast catching the only bird species that is restricted to South Australia plus an elusive grasswren in one net! We had been so captivated with the whitefaces that it was nearly four o'clock before we realised that we probably deserved lunch. It is amazing how food cravings are inversely proportional to the amount of fun you are having. Perhaps Jenny Craig should incorporate studies of chestnut-breasted whitefaces into her weight-loss packages!

But I have digressed. I was making the point that the *Bird Atlas* is inherently biased towards rare species or notable sites and I got carried away with the chestnut-breasted whiteface story. That is because under the 'Birds Not Listed' section of the pink form, I invariably see the chestnut-breasted whiteface recorded by most birdos travelling up the Strzelecki Track. Given its incredibly restricted distribution these records are ample proof that although the *Atlas* can successfully map bird distributions it fails to representatively survey bird abundance.

Although the sites selected for birdwatching are obviously biased, I am firmly convinced that the *Bird Atlas* is a worthy project. Birding is a better pursuit if you can share your experiences with others. Anyone who is interested in birds is obviously going to make a special effort to try to locate rare or spectacular species. I remember having an ordinary time on a holiday in the Simpson Desert until I saw an Eyrean grasswren, which is an elusive bird largely restricted to the desert. Often it is the quest for the unusual that keeps birdos enthused and without the enthusiasm of amateurs our knowledge of bird distribution and ecology would be considerably more rudimentary.

✸ ✸ ✸

A buzz of excitement jolted my gaze skywards. The high-pitched alarm calls from the white-plumed and grey-fronted honeyeaters were racing like a wave towards me as I craned my neck to find the cause of their concern. What would I see gliding through the low canopy of eucalypts in search of an inexperienced, injured or distracted bird that had not responded to the alarm? Surely the black kites circling overhead had not stirred up the honeyeaters. The threat would probably be a sparrowhawk, goshawk or little falcon, but even owls or butcherbirds can elicit such a response from the local honeyeaters. The wave passed me without revealing its cause, which I suspect was hidden behind my neighbour's house. For I was not bird-watching in mallee scrub. Nor was I atlassing in a remote eucalyptus hot spot. Rather I was hanging up the washing in my yard at Roxby.

When I first came to Roxby, the place where I was now standing was a low red sandhill with scattered white cypress pines and sandhill wattles. There were no fences or houses to obscure my vision of an approaching raptor. But back then there may not have been a bird of prey zipping past and there definitely would not have been any white-plumed honeyeaters to alert me if there was. Instead I might have seen a flock of thornbills, variegated fairy-wrens and some black-faced woodswallows gliding above the trees. In the past fifteen years the landscape has changed from a sparse arid scrubland to a eucalyptus woodland. And the bird life has changed too.

Pardalotes were now calling from the canopies of gum trees when thornbills had once flitted between old pines. White-breasted woodswallows had largely replaced their inappropriately named black-faced cousins. Introduced sparrows had appeared where zebra finches once dominated, and black-faced cuckoo-shrikes had taken over from ground cuckoo-shrikes. Corellas now slide raucously down the white conical roof of the motor inn and galahs spin frivolously on powerlines where the less extravagant bluebonnet parrots once winged their way through open woodland. Welcome swallows and martins now swarm in the evening sky and rainbow bee-eaters congregate excitedly in their favourite roost trees in town. The bee-eaters particularly enjoyed the town in the mid-1990s when a hive of honeybees was imported for the benefit of a local multiple sclerosis sufferer. Now they make do on the buzz of native insects that accumulate around flowering trees and shrubs, in one of the few Australian towns without feral bees or wasps. On occasions, birds such as oriental cuckoos, blackbirds and white-throated gerygones, which are normally not found within hundreds or thousands of kilometres of Roxby, also turn up in this modified, greener environment.

The current *Bird Atlas* will reflect these changes. Equally as important as documenting colonisation events is recording the decline of species. The *Bird Atlas* is a powerful tool for identifying contracting bird distributions. An even better technique is the long-term, regular recording of bird species from a particular site. A classic example of such has been the eighteen-year Aldinga Scrub bird study by local ornithologist, Colin Ashton.

The Aldinga Scrub is a square patch of coastal woodland that can be seen within the agricultural desert and urban sprawl on the drive to the surf beaches south of Adelaide. In the 1970s Colin recorded scarlet robins, diamond firetails, white-browed babblers and jacky winters that he has not recorded in the 275-hectare conservation park since 1990. Many other birds also seem to be gradually disappearing from Aldinga Scrub. After 400 visits Colin has found that the number of superb fairy-wren flocks has declined from nine to three, hooded robins have declined from ten pairs to three and the numbers of yellow-rumped thornbills have halved.

This disappearance of birds from Aldinga Scrub is just one well-studied example of much broader and alarming trends. The South Australian Ornithological Society has documented the serious demise or loss of at least forty-two bird species in the Adelaide and Mt Lofty Ranges region in the past few decades. Even though clearing of native vegetation, the main cause of their decline, ceased a few decades ago, birds are still disappearing from 'islands' of scrub. David Paton warns that half of the Mt Lofty Ranges species introduced to me during my uni days could disappear if these trends continue.

Many birds also live in habitat 'islands' which are analogous to the isolated patches of land suitable for Pernatty knobtails or plains rats. The splendid fairy-wren could well be named because it is a splendid example of a 'habitat island' dweller. I know of three isolated populations of these outrageously beautiful blue birds within thirty kilometres of Roxby Downs. They only live in particularly dense patches of dune vegetation and are replaced by variegated fairy-wrens on most other dunes. The incredible site fidelity of these tiny birds was revealed when I retrapped a male eight years after its first capture in exactly the same location. It is highly probable that for all that time he did not leave his densely vegetated sand dune 'island'.

Loss of biodiversity occurs wherever natural habitat islands are degraded or when developments fragment widescale habitats into islands surrounded by hostile seas. Declines in wildlife diversity will continue to occur wherever we limit the quality or the size of the remaining habitat islands. As they say in the Pantene jingle, 'It won't happen overnight, but it will happen.' Because threatening processes such as predation, disease, habitat change and dietary stress operate more quickly and with greater affect on islands than they do in large areas, birds like the splendid fairy-wren are especially vulnerable to changes. The *Atlas* is probably the best way of revealing changes to the viability of small island populations of birds especially around the periphery of their range. These may also provide an early indication of detrimental impacts elsewhere.

The success and usefulness of the original *Bird Atlas* was probably a major catalyst for Frogwatch, Batwatch and an increasing array of national and regional biological surveys. Ideally we would also have atlases or 'watches' for lizards, beetles, ants, plants, lichens, fungi and the list goes on. Changes in the abundance or distribution of these organisms will highlight other consequences of our tampering with the environment. However, it is unlikely that any of these other initiatives will be able to compete with the *Bird Atlas* in the number of volunteers that can be co-opted and the range of sites that can be covered. Nevertheless, because bird populations are closely linked with environmental conditions they are a useful surrogate for monitoring changes in the environment in general.

By spending weekends and holidays looking for elusive chestnut-breasted whitefaces, rare waders or spunky splendid fairy-wrens, *Bird Atlas* volunteers are contributing to one of the most effective 'State of the Environment' surveys possible. The concern of many ecologists is that by the time birds start disappearing it may be too late for us to remedy their decline. We also need to monitor more sensitive barometers of change and consider the long-term repercussions of any tampering we do with our already embattled environment.

17
Buckets of frogs

My most vivid environmental-monitoring memory is of checking for abnormalities in frogs collected by Filipino villagers from streams adjacent to a potential mine site. After sorting through several baskets of frogs with no deformities I was intrigued that one receptacle, fashioned from a plastic drink bottle, contained several frogs with fused or missing toes. When I double-checked with the bloke who collected them I was dumbstruck.

A middle-aged local stepped forwards from the throng and informed me that he had collected the frogs from Alutayan Creek. What startled me was that he only had four toes. When I regained my composure I learned that the collector was named Sinyalan—a reference to his missing toe that was considered a symbol of good luck. Sinyalan was born adjacent to the same creek where he had collected the frogs.

As the morning progressed a trend in the abnormalities became apparent. Most of the 'abnormal' frogs had been collected from Alutayan Creek. We found frogs missing complete toes and others with shortened, fused or missing bones. This milky looking creek contained naturally high levels of copper, mercury and nickel as well as being extremely acidic. Trying not to offend I asked if there were other local people with good luck symbols, and I was informed that it was not uncommon. Although I cannot confirm that Sinyalan's abnormality was linked to the toxic water, the cocktail of contaminants would have undoubtedly contributed to the deformities in the local frogs.

Frogs are such good indicators of water quality due to their unique development process. Unlike most other vertebrates whose eggs are fertilised and develop within the relative protection of the mother's body or in a largely impervious egg, amphibians' eggs are fertilised and develop in water. Therefore, even before conception and particularly during their rapid growth period, the developing amphibian may be exposed to aquatic contaminants that readily pass through the pervious egg. But that is only part of the story.

Unlike other vertebrates, which are born or hatch with their full complement of bones, frogs go through the insect-like phase of being a larva, or tadpole. These aquatic larvae live in streams or ponds where pollutants accumulate after being washed from the land. Tadpoles are effectively swimming vacuum cleaners for toxic chemicals. It is only after this dual exposure to contaminants while in the form of an egg and a tadpole that a frog undergoes metamorphosis, that amazing change that has thrilled endless generations of kids and scientists. Only at this stage do tadpoles, packed with stored energy and a possibly toxic legacy, develop their skeleton in preparation for adult life. It is little wonder that frogs that develop in contaminated ponds or streams often end up a bit peculiar.

Around the world deformed frogs are synonymous with contaminated environments. Over a third of the frogs sampled from ponds contaminated by Russian municipal wastes and sewerage were seriously malformed. So too were about half of the frogs collected from a contaminated quarry site near Stuttgart. Twelve percent of the frogs sampled from Canadian farmlands exposed to pesticide-infused run-off exhibited severe deformities. Even more alarming is that Spanish researchers found that practically all of the frogs exposed to commonly used pesticides as tadpoles had deformities that would have seriously affected their long-term survival. Closer to home, Mike

Tyler from Adelaide University found that over a third of the frogs from the naturally radioactive Paralana Springs in the Flinders Ranges were deformed in some way. Therefore, if our environments are contaminated, there is a good chance that high numbers of deformed frogs will turn up.

※ ※ ※

Olympic Dam uses and produces a range of chemicals, elements and by-products including radiation that could affect the environment and particularly the frogs. So, what's going on? Have I found deformed frogs at Roxby? Yes I have. In fact about the tenth frog I ever looked at near the mine sported an extra arm. Oh my goodness! This was a symbol I thought, a very bad symbol. I got straight on the blower to Mike Tyler, the same bloke who had studied the deformed frogs in the Flinders Ranges.

Mike is a lanky eccentric scientist who can only be described as a 'frog freak'. He has dabbled in more ponds and more aspects of frog biology than any other Australian in history. His illustrious frogging career has given him the distinction of describing over sixty living and fossil frog species, and yielded nearly 400 scientific publications. Mike is an Associate Professor of Zoology at Adelaide University, ex-director of the South Australian Museum, and an expert in attracting attention from the media and funding bodies. As a result, he always seemed to have a bemused attachment of students trying to decipher and cope with his bizarre mannerisms. With his reading glasses perched erratically on his nose and his habit of excitedly pacing back and forwards and orating comically to the walls and ceiling, he had the knack of entertaining, informing and confusing me at the same time.

Mike had studied the frogs at Olympic Dam as part of the Environmental Impact Statement in 1981. He told me that my five-legged frog was interesting, alarming and meaningless. I was to call him back when I had done a 'decent' survey and was told to hang on to any frogs that 'departed from bilateral symmetry'. Apparently it was perfectly natural for one to five percent of the frogs in an uncontaminated area to exhibit abnormalities. I wouldn't be able to tell if the frogs at Roxby were sending a signal until I had looked at hundreds or even thousands of them. So that's exactly what I did.

※ ※ ※

The Pitgrid is one hectare of saltbush scrub about a twenty-minute drive north of Roxby. It is a series of 401 half-metre deep lengths of sewer pipe, dug into the rocky clay soil at five-metre intervals. It took a couple of months and a rack of blisters to dig all of the holes in 1990. Since then, and with the assistance of enthusiasts who submit to centipede bites and scorpion stings, I have trapped the Pitgrid for more than 200 nights over a decade. And it has been worth every minute and every sting.

Through the Pitgrid I have been able to follow the movements, habitat preferences and typical lifespan of thirty-five reptile and seven mammal species. That is more types of lizard in one hectare of desert scrub than in all of New Zealand and the United Kingdom combined. Of most interest are the population changes through time. Some species such as house mice or beaded geckos are not trapped for a year or more and then turn up in dozens, or even hundreds. In some years the skink *Ctenotus leonhardii* was scarcely recorded, in others it was almost the most common reptile sampled—a bizarre occurrence since some individuals of this species lived in the Pitgrid for at least seven years. However, the most striking information to come out of the Pitgrid was the abundance of frogs. I had thought that Mike Tyler was asking for the impossible when he suggested that I look at thousands of frogs in one of Australia's driest regions. It turned out to be a cinch.

The most abundant lizards in the Pitgrid are a skink called *Ctenotus regius* and the little beaked gecko, *Rynchoedura ornata*. Over the past decade more than 350 of each these lizards have been captured in this hectare. The maximum daily tally for both of these common lizards stands at eighteen individuals. However, in February 1997, 163 trilling frogs were captured in one day and in 1992 the peak daily tally was 145. This trapping alone suggests that at this site frogs are up to nine times more prolific than the most abundant lizard. These figures are even more staggering when you consider that the Pitgrid is not near any swamps and that the pits are not fenced to guide frogs in from elsewhere. When traps are placed near swamps the number of frogs captured is even more staggering.

In May 1989 I was slipping through the mud and closing pit traps that had been inundated by an overnight deluge. Many traps had yielded a dozen or so frogs, but I was not expecting what I found when I crouched down at a pit at the base of a dune near a small swamp. Rather than bumping into a few frogs as I plunged my hand into the cold murky water, it felt like I was lucky-dipping into a pipe full of slimy rubber balls. Each time I put my hand in it was met with a slippery, squeaking resistance as I extricated another handful of squirming frog-flesh. In due course 244 frogs were emptied from the single 150-millimetre diameter pit.

All of these frogs were trilling frogs, the only species that has been recorded at Roxby. They are dumpy greenish brown frogs that grow to about the size of a golf ball. Although all the frogs were the same type they came in a bewildering array of camouflage patterns of grey, brown and grey–green blotches. Some even sported a narrow pink pinstripe down their spine.

On the way back to my laboratory I stopped at a swamp in which I had been monitoring the development of tadpoles since the huge March rains. The first metamorphlings had left the flooded saltbush pond about a week earlier, but there were still hundreds of taddies at different stages preparing for life on the land. Unfortunately, many had taken their first and last steps. Only minutes earlier a 4WD had driven around the swamp and in just thirty metres of tracks from a single car, 281 juvenile frogs had been squashed. Armies of little hoppers swarmed past their fallen comrades. Anyone who has not witnessed these swarms of frogs in the desert after rain finds it difficult to comprehend.

Fortunately, more people are witnessing this phenomenon in a variety of places. In March 1996 I was fortunate enough to be in the Great Sandy Desert, just south of the Kimberleys in Western Australia, when Cyclone Kirsty dumped nine inches of rain on the spinifex desert in one day. As soon as the waterfall of rain abated I headed out with Darren Niejalke to check out the response. We had far more success than our quest for rare mammals at Lake Eyre. Among the flooded bilby holes and drowned marsupial moles we saw hundreds of thousands of frogs. Unlike at Roxby, the Great Sandy Desert supported at least five frog species. More frogs can live in this beautiful spinifex desert because the big rains, like the one we had just experienced, occur more regularly than in the super-dry regions like Roxby. The trilling frog is the most arid-adapted frog in Australia. It is the only species that can reproduce quickly enough to breed in temporary ponds and also survive the long dry periods between rains to live in the driest parts of the continent.

The most abundant of the frogs in the Great Sandy Desert was the desert spadefoot toad—a relatively large frog that exudes a white sticky goo resembling Aquadere when handled. This round, toad-like burrowing frog with multicoloured raised bumps is also super-abundant in other sandy deserts. Chris Dickman, who heads up the University of Sydney ecological laboratory, has conducted extensive studies of the wildlife of the Simpson Desert. Chris was fortunate enough to be caught in a similar deluge of rain and frogs. Desert spadefoots were also the dominant terrestrial vertebrates at his study site, both in terms of biomass and sheer numbers. With co-worker Martin Predavec, Chris determined that the combined biomass of three species of frogs at their study site was a staggering 2,315 grams per hectare. Even more staggering is that these frogs are not restricted to the margins of ponds or temporary creeks. Steve Morton, the refugia man, and some other scientists were amazed to find up to sixty-eight frogs per hectare in a spinifex sandplain of the Tanami desert—that was at least ten kilometres from any pond that held water for more than a day or so.

This abundance of frogs in Australian deserts has only recently been recognised by scientists in what has traditionally been regarded as the 'land of the lizards'. Neither mammals nor birds can match the adaptations of lizards, which are physiologically and ecologically tuned to living in hot dry environments with unpredictable rain and predominantly small insect prey. If the birds and mammals can't cope so well, how can frogs be so spectacularly successful in an environment that is so obviously suited to reptiles?

Unlike some insectivorous mammals and birds that may find themselves in direct competition with reptiles for the few insects that are available during dry times, frogs totally avoid this battle. In fact most desert frogs simply avoid facing up to dry times altogether. Before the rain-softened desert earth bakes hard again, frogs burrow deep into the moist soil at the edge of ponds, under bushes or on damp sand. They then form their own watertight cocoon from successive layers of exfoliated skin. Within this cocoon and with their metabolism wound back like hibernating bears they can sit out months or even years until rain softens the soil and they can re-emerge. Trilling frogs at Roxby are active for an average of twenty nights a year, although in a two-and-a-half-year drought in the mid-1980s most were probably active for a total of only eight nights.

Those nights when buckets of frogs are active after rain are the nights when ants, termites and other insects are swarming in such huge numbers that there is plenty to share among the animals that dare to brave the wet conditions. Another result of these long periods of inactivity is that frogs mature slowly and live for a long time. It is little wonder that they live so long when you consider that an eighteen-year-old frog has probably only been active for about a year in its whole life. A long life span is essential for trilling frogs since they only have the opportunity to breed two or three times a decade.

※ ※ ※

So frogs are good indicators of environmental health, they are locally superabundant and deformed frogs have been found at Roxby. Given all of these factors, what do the local frogs tell us about the environment around the mine? Unlike those poor frogs living near toxic ponds in Europe, most of the trilling frogs in the Roxby area are apparently 'normal' and do not exhibit any obvious abnormalities. Of course, some frogs collected near the mine and other 'control' frogs collected tens of kilometres away are deformed in some manner. In each of the three surveys conducted after the 1989, 1992 and 1997 rains, abnormality rates both near and remote from the mine have been less than four percent. This seems like an awful lot, but we have to remember that we are dealing with frogs here. As anyone who has ever bred a large numbers of frogs will tell you, four slightly deformed individuals out of 100 is a pretty clean score.

It is also important to remember that not all deformed frogs have necessarily been affected by contaminants. Injuries, particularly old healed ones, can resemble skeletal abnormalities. Parasites or common genetic mutants can also increase abnormality levels in perfectly clean environments. One in twenty Australian arid zone frogs from unpolluted environments have some deformities. In contrast, usually at least one in ten frogs from contaminated sites exhibit skeletal deformities. Therefore Mike Tyler and I concluded that the local environment was relatively clean.

Although frogs are superabundant and supersensitive to pollutants, they are not ideal animals for monitoring the environmental impacts at Olympic Dam for three reasons. Their long life and mobility suggest that a frog inspected near the mine site may have either metamorphosed before the mine started production, or may have hopped in from a remote pond. Because the abnormalities of interest only occur during metamorphosis, we could potentially be looking at healthy-looking frogs living in a contaminated environment.

The second main drawback is that fatal deformities may be occurring in the frogs that would not be registered in our surveys. Even if frogs from contaminated ponds are not deformed they may be less healthy due to changes in their physiology or metabolism. A little frog with a twisted back foot or a few missing toes will probably not be able to screw itself into the dirt as efficiently as a healthy frog. Deep burial is vital to their survival and hence deformed frogs may die soon after metamorphosis. We will not be able to reveal an increased percentage of deformed frogs if the affected individuals have died!

Both of these shortcomings can be overcome by sampling metamorphlings as they emerge from ponds. That is what we did following the last main breeding event in 1997, and fortunately the young frogs still gave the mine the thumbs up. However, the most obvious drawback with using burrowing frogs as sensitive indicators of the 'health' of the environment is that they are seldom around. Trilling frogs can't be sampled at Roxby unless we have big rains. A check-up every five or so years is not frequent enough for environmental monitoring. We need to find critters that can be easily and regularly sampled that are also sensitive to pollutants. That was my main quest during work hours for most of the 1990s.

18
Outback canaries

When I was a kid I used to fantasise about talking to animals, like my one-time hero Dr Doolittle. Occasionally I still consider the possibilities. Wouldn't it be great to have a yarn with an eagle, a goanna or a dolphin and learn how they see the world? What a fantastic recruit Dr Doolittle would be for a wildlife management group, a research organisation or a mining company. Direct communication with animals would cut-out our guesswork as to how to solve their problems. We could then address the real environmental issues as perceived by the critters that we were 'looking after'.

In a slightly less fantastic way the mining industry has been 'talking' with animals for centuries. Canaries are the best-known examples. If a caged canary fell off its perch, coal miners knew that the amounts of odourless carbon monoxide had increased to dangerous levels. We used, or abused, canaries for our own benefit. Fancy machines and well-trained staff now look after the health of the miners at Roxby, while I focus on the environmental issues that affect the animals' lives.

I needed to find the equivalent of caged canaries at Roxby, but I wanted much more than just that. Canaries are sensitive to carbon monoxide but are not necessarily affected rapidly by other environmental insults. I wanted to find 'canaries' that were sensitive to all of the types of pollutants that a mine may produce, including the sulphur gasses from smelting copper, salt spray from underground ventilation shafts, radiation, noise and dust. An obvious starting point was to investigate the wild animals that lived in the area to see whether they could help me out.

Logic suggests that a species that had declined since European settlement would be sensitive to disturbance and hence would be a good indicator of the environmental condition. With the simplistic hype surrounding the concept of 'biodiversity conservation', losing rare species is thought to be worse than causing declines in common species. Therefore, I figured that monitoring efforts should initially concentrate on rare species, although I quickly learnt that these animals are not necessarily good canaries. Three of the rarest animals that I could have remotely expected to record at Roxby—the plains wanderer, bustard and the desert mouse—all proved to be at least as common in areas disturbed by mining as in the more pristine areas. Indeed, the desert mouse, which had not been recorded in the State for nineteen years, seems to actually favour the noisy, salt-sprayed sites near the mine.

Not only may endangered or rare species be insensitive to some environmental changes, it is often difficult to find enough of them to reach definite conclusions. A far more profitable approach is to investigate some of the more common animals to see if there are any canaries among them.

First I looked at reptiles. I have always had a particular interest in snakes and lizards and loved the idea of catching vast numbers of them under the guise of work. It also made sense. Lizards were both abundant and diverse at Roxby with over fifty species recorded from the local region. Not only are they fun to catch, lizards are easy to identify and many of the common species only live for a couple of years. Short life spans mean that their populations may change rapidly if they run into problems. What was not known was whether they were sensitive to disturbance or pollutants.

The first species to whistle like canaries were netted dragons. These orange-flushed, mushroom-coloured lizards with intricate dark-brown markings that resemble netting, grow to about a foot in length in less than a year. Perhaps more than any other local species netted dragons responded dramatically and rapidly to disturbance. Also somewhat counter-intuitively they seemed to follow bulldozers. Netted dragons rapidly colonised areas that had been cleared for the development of Roxby Downs, the construction of the processing plant and along the edges of newly graded roads. On one day in January 1991, Rachel Paltridge, who was working for me on her university vacation, counted forty-five netted dragons basking on tree guards on the three-kilometre Eagle Freeway to the mine.

If you drive along the same road in summer these days you will be lucky to see four netted dragons. As the vegetation recolonised the road verge these sun-loving lizards returned to approximately the same population densities as before the roadway was cleared. In the same way, netted dragons are now rarely seen in the Roxby Downs town, except for when a new street is being added to the outskirts. Where hundreds of dragons roamed the exposed red sand fifteen years ago, garden skinks now scuttle through the ground cover and leaf-litter of maturing gardens. Although netted dragons are great indicators of scrub that has been razed by bulldozers, cattle or fire, they only indicate the bleeding obvious. No-one needs a dragon to tell them that a bulldozer or herd of cattle had cleared the vegetation. I needed indicators of more subtle, pollution-related changes.

In terms of their shape, habits and physiology, the local geckos are the antithesis of netted dragons. Geckos are generally small, slow and active by night not day. Whereas netted dragons eat mainly leaves and the females lay many eggs in a clutch, the local geckos solely eat insects and lay a maximum of two eggs per clutch. With tough skin and small eyes dragons are built to withstand dust storms in the height of summer. Geckos have delicate soft skin and large eyes that they lick with their tongues because they have no eyelids. If dragons thrive in disturbed country one would expect geckos to suffer, and they do.

Because the Olympic Dam mine is underground, land disturbance is contained within a relatively small area. The most pervasive environmental effects of the mine and processing plant are from airborne pollution. Sulphur dioxide, a waste product of smelting copper, was the main pollutant originating from the Olympic Dam mine in the early 1990s. Sulphur dioxide and allied gases are the toxic ingredient of much of the world's acid rain, which destroys environments adjacent to industrialised regions such as Germany's Black Forest. Although more than ninety-five percent of these gases at Olympic Dam were converted to sulphuric acid I had an inkling that the remainder may still affect the surrounding environment. Likewise, in localised zones around the mine ventilation shafts I could often taste the salt spray in the air that had been forced upwards through a saline aquifer. If I could taste the salt, I bet the geckos could too.

An interesting pattern emerged after catching hundreds of geckos from sites near the smelter and ventilation shafts, as well as hundreds more at sites remote from any disturbance. Firstly, as expected I caught considerably fewer geckos near the mine than at control sites. The poor little critters obviously did not appreciate the contact with foul-tasting sulphur gases or salt spray, especially when they licked their large eyes to remove the irritation. But there was more to it than that.

As anyone who has tried to breed animals knows: unstressed, healthy individuals usually produce the most offspring. Because the belly wall of geckos is relatively transparent, it is easy to see if they have eggs. Nearly all of the adult female geckos that I caught away from the mine had eggs. Herpetologists say that they were 'gravid'—the egg equivalent of being pregnant. Because they laid successive clutches throughout the summer, most unstressed geckos were almost constantly gravid during the warmer months when I did my sampling. By contrast only about half of the females near the smelter and ventilation bores were gravid. Something was wrong.

The two species that were most affected—the beaked and fat-tailed geckos—feed exclusively on termites. Species such as crowned geckos and Bynoe's geckos that feed on a wide range of insects suffered far less. They were not as sensitive. Maybe termites were affected by air pollution.

Although not as conspicuous as in the spinifex deserts or speargrass savannas of northern Australia, termites are incredibly abundant and important insects at Roxby. During a dry period several of the common grasses remain as chewed-down pincushions of dry stalks. At first I thought that rabbits or kangaroos were to blame for crew-cutting the grasses, but careful inspection on humid nights revealed the real culprits. Millions of termites seek out these dried grasses and like miniature woodchoppers they chew off segments five to fifteen millimetres in length, and carry them back to their underground galleries. Many common lizards and invertebrates, as well as trilling frogs, feed primarily upon termites. Therefore, declines in termite numbers around the mine could have serious consequences for many other animals. They could be the ultimate 'canaries' that I was after.

The difficulty with monitoring termite numbers on saltbush flats is that they are difficult to locate. Their nests are invariably underground with openings that are sealed when the termites are not active on the surface. Because the local species mainly feed on dried bluebush leaves or grass tussocks they are not particularly attracted to typical termite baits like buried toilet rolls or wood. I reckon that the easiest, and definitely the most fun way of monitoring changes in termite numbers is to monitor changes in termite-specialist gecko numbers. The Olympic Dam engineers gave me a perfect opportunity to do just that.

Partially in response to our initial environmental findings, the ground concentrations of both sulphur dioxide and salt spray were reduced in the early 1990s through modifications to both the smelter and the ventilation bores. Sure enough, the next year when I surveyed the wildlife, more geckos in both impact zones were gravid and gecko capture rates had also increased. These findings suggested that geckos might be excellent indicators of the health of the environment and a decade later they are still used as little outback canaries at Roxby.

Unfortunately, geckos can't tell us everything about the condition of the environment around the mine. For a start, it takes a day or so to dig the thirteen pit traps and ten consecutive days to trap a meaningful number of geckos for any one site. Dozens of ecologists would be required to sample the entire 180 square kilometre mine lease on warm summer nights if we relied on geckos alone. Furthermore, geckos might only be indicating where the air pollution was affecting them or their termite prey. How about the environmental impacts of noise, dust and radiation? I needed to find other 'canaries' that could be monitored for different impacts at a broader range of sites.

Ants were an obvious choice. They are by far the most conspicuous, widespread and arguably the most important animals in the Australian arid zone. Ants are superabundant, voracious predators of seeds and insects, valuable seed dispersers, scavengers and important prey for many lizards. Because of their huge diversity and range of ecological roles, ants are sure to respond to most changes in the environment.

Ants should be our ultimate canaries, but they suffer terribly from an image problem. Although anyone who has looked at a *Calomyrmex* through a microscope would disagree emphatically, ants are seldom considered to be sexy, cute or valuable. They are just too small and often too difficult to identify to attract the interest of the general public and much of the scientific community. As a result we know little about our ants. In fact we probably know more about the biology, distribution, habitat preferences and principal threats to our most poorly understood bird species than we do for our best-known ant species. Most of our ants do not even have names. How can we use them as bioindicators if we don't even know their name or what makes them tick?

This situation is changing; Australian scientists are gradually learning about key ant species and how they respond to environmental change. Some ants, like the little pissants that take over the gardens and houses in Roxby or the larger meat ants that turn up at every bbq site or dump, obviously thrive in areas that have been altered from their natural state. Because these annoying species are typically aggressive and common, they displace some of the more submissive or less common species. By looking at the ratio of ants that are favoured by disturbance versus those that shy away from these regions, we can develop an understanding of how 'pristine' a location is. With such numerous categories of ants there is still a fair amount of uncertainty about which category the ants belong to when we are investigating different types of environmental disturbance. However, I reckon there is a fair chance that ants will eventually prove to be as valuable as frogs and geckos combined in signalling the condition of the environment.

※ ※ ※

Identifying particular animals and plants that can act as canaries is important, but it is only the start of the story. A mine manager will at best fix me with a quizzical stare if I announce that gravid beaked geckos or *Rhytidoponera* ant numbers have changed on a particular dune near the mine. The responses of each of our canaries needs to be interpreted to give an overall picture of the health of the environment. This sounds complicated but there is a simple way around it. In the same way that surveyors project contour lines from spot readings of elevation and meteorologists construct isohyets from a scattering of rainfall gauges, industrial ecologists need to combine the response of all of our different canaries by drawing lines around areas of equivalent environmental health. These lines are called biohyets. Someday if 'biohyet' gets into the dictionary it will probably be defined as 'a line of equivalent biological integrity', although we know it simply shows us the volume at which the canaries are singing!

Not surprisingly, another group of animals being used to construct biohyets at Roxby is birds. After several years of monitoring that was more intensive and targeted than the Bird's Australia *Atlas* surveys, some discernible trends became evident. A few bird species were consistently recorded more frequently away from the mine than nearby. The crested bellbird was one of them. Unlike their gregarious, aggressive namesakes from eucalypt forests, the bellbirds in the desert are largely solitary and usually shy. Crested bellbirds can often be heard in the morning chiming their distinctive ventriloquial call from the top of a shrub. More rarely they can be seen jumping around on the ground under tall bushes with their little crest raised. However, near the mine I seldom heard or saw them. Maybe they could not compete with the clattering, beeping and rumbling noises from all of the machinery.

The other birds that seemed to opt for the quiet life were among the smallest birds in the country. Thornbills are tiny insect-eating feather balls that fly around in flocks with several other bird species. It is not uncommon to wander through wattle scrub for ten minutes without seeing or hearing many birds, then all of a sudden be confronted with a twittering explosion of activity. Inland thornbills, chestnut-rumped thornbills, southern whitefaces, variegated fairy-wrens and perhaps the odd yellow-rumped thornbill, red-capped robin or woodswallow can all group together, excitedly squeaking out their alarm calls. The thornbills revel in these mixed flocks, imitating the calls of their flock mates. As a result, to work out which species are present, you have to check out each individual of the continuously moving flock to establish if they have brown or white eyes and chestnut, grey or yellow-coloured rumps.

These mixed-species flocks are not nearly as common near the mine as they are on remote dunes. Maybe the insects that they eat are less abundant, maybe the industrial noises interfere with their calls, or maybe one or more of the bird species simply don't like the smells, tastes or dust associated with the mine. Whatever the reason they appear to be good indicators, and along with crested bellbirds are used to map environmental health at Roxby.

The remainder of the diverse bird assemblage does not help with identifying environmental impacts at Roxby. Most of the birds recorded from the region are either nomadic or otherwise highly mobile. A flock of budgies, chats or trillers will indicate recent rain but not whether the mine has affected the surrounding country. Of those resident birds that are potentially useful, most seem to be at best ambivalent to the noises, smells and structures associated with the mine. In fact most birds that have responded to the mine have increased in numbers.

Finches and pigeons thrive on the permanent drinking water available from water storage and sewerage ponds. Magpie-larks, white-breasted woodswallows and fairy martins make use of the mud and buildings for nesting. Kestrels and kites soar on the updrafts from the rock dumps, while kingfishers and white-backed swallows nest in quarry faces or road cuttings. Several other bird species including wedgies, honeyeaters, pipits and fairy-wrens also raise apparently healthy clutches of chicks in the immediate vicinity of the mining operation, which suggests that the mine has had little or no effect on them. The most extreme case of birds nesting in a seemingly dodgy location involved a clutch of red-backed kingfishers.

Despite their name and the habits of many of their close relatives, red-backed kingfishers neither live near water nor do they feed on fish. They live in the desert, feed on insects and lizards and nest in holes that they excavate in vertical banks. One particular pair of kingfishers decided to construct their nesting burrow in the side of a stockpile of ore at the mine. This was not any dusty stockpile of crushed rock. The area that the kingfishers chose was a pile of ore that was particularly rich in uranium. Different grades of ore are stockpiled separately and then combined to achieve the desired mix of copper, gold, silver and uranium for the processing plant. The kingfishers just happened to pick the 'hottest' stockpile for their burrow, and the precise location that would ensure their chicks received the maximum dose of radiation at Roxby.

A loader operator rang to say that he had unknowingly half

obliterated the nesting tunnel while retrieving ore. I was told that if I got there soon I might be able to save the nearly fledged chicks. The sight that greeted me was appalling. One half-naked chick had fallen from its exposed nesting chamber and was floundering around in the full sun. The fine purple dust from the ore had caked into its beak and eyes while its parents looked on helplessly from a nearby tree. Its nest mate had remained in the burrow, but was stained purple from the ore and was also exposed to the sun. I had no option but to take the chicks away to wash and rehydrate them.

As I left the area it dawned on me that these two dishevelled birds had been exposed to more radiation than any of the underground miners or local wildlife would in a lifetime. Gamma rays had passed through the eggshell and skin of the hatched chicks. Radon gas permeated through the eggshell and was inhaled by the chicks after hatching. Radioactive dust was ingested with food items. The effects of low levels of radiation are greatest in developing embryos and hence these chicks had been exposed to elevated radiation levels during their most sensitive, formative weeks. What I found even more surprising was that they appeared to be healthy. Indeed, after I fed them for a few days they were taken in by an Adelaide fauna centre where they successfully fledged and lived for many years, apparently unharmed by their eventful and seemingly hazardous upbringing.

The survival of these kingfishers and the increase in bird numbers around the mine does not indicate that the mine has been 'beneficial' for the local bird community. It simply means that some birds, along with netted dragons and desert mice, have benefited from the changes to their habitat and the provision of water, food or nest sites—probably at the expense of other animal and plant species.

The continued successful breeding by many bird, reptile and mammal species near the mine site, together with the consistent recording of low levels of abnormalities in metamorphling frogs, does indicate something that is significant. While sulphur gases, salt spray and ground disturbance do affect some local animals, most species appear to be either unaware of or unaffected by the slightly elevated radiation levels found around the mine. I had answered, in my own mind, one of the most controversial questions that I set out to solve when I first came to Roxby.

PART FIVE
Whose outback is it?

19
Who's shittin' who?

'It's heavy shit man, you've got to listen to me brotha 'cos what I'm goin' to warn you about will save your life man.' Although I was dubious, I listened anyway. 'You know about meteorites ... you know what a meteorite is don't you? See there's going to be this meteorite man and I'll, I'll draw you a map.'

They called him Mad Mark and he bounced away to get a gibber stone before excitedly rushing back and wiping clear a patch of dirt at my feet. The low rise above Lake Eyre South where the anti-nuclear activists had set up their protest camp was his stage, and he was loving it.

Mad Mark did not strike me as the sort of bloke who had a handle on the predicted trajectory of meteorites. He was a rangy mangy creature whose hair and beard seemed to be waging a battle to develop the best dreadlocks. With the aid of his rock and dust patch and enhanced by flailing arms and freaky eyes Mad Mark explained how the giant meteorite was going to land in Spencer Gulf near Port Augusta. The huge resultant tidal wave was going to inundate the north of South Australia and this inland sea was going to drain into the underground Olympic Dam mine. The whole area would then be contaminated in 'deadly shit'. I was in danger man and I had to leave Roxby. It was most fortunate that I apparently resembled Mad Mark's brother. Otherwise I would not be privy to the top secret maps he was drawing for me of a hideaway north of Byron Bay where I would be safe.

It was not Mad Mark whom I had come to visit. The bloke who I really wanted to catch up with was sneaking around the back of my car. I was watching him out of the corner of my eye making sure that he was doing nothing more than placing a few stickers in discrete locations that he hoped I would not notice.

'I've brought up some information on the mound springs ... Are you keen to talk about it?' I asked, beckoning him over.

'Johnny Rambo' did not seem particularly interested in listening. He was clearly on a mission and talking with the enemy was not in his manifesto. He was dressed in the same army fatigues that he was wearing when we first met the other week. I had been driving back from the William Creek races with some mates and we had pulled up next to the Aboriginal flags to hear what the protestors had to say. Johnny Rambo barked our number plate and description into one of the several CB radios he was carrying. He then informed us that they were the Keepers of Lake Eyre and that they were protecting the lake and mound springs from WMC on behalf of the Arabunna people. The fact that the waters of Lake Eyre and the mound springs were in no way connected, and just happened to be found in the same region through geographical accident, was lost on Johnny Rambo. The absence of the local Arabunna community from their campsite was also intriguing.

With the innocence of anonymity we asked what WMC was doing to Lake Eyre and the mound springs. We were told that radiation had seeped from the tailings dam into the Great Artesian Basin and that the mining company had displaced the traditional owners from their land. What's more, the Bubbler mound spring used to fountain water three feet into the air but now, since the mining company was taking water, the bubbles barely broke the surface.

It was too much. I asked him who had told him that the Bubbler used to spurt out a three foot geyser of water and when this had occurred. He said that his group had witnessed this event within the past ten years. I blew our anonymity and said that I had been regularly visiting the Bubbler for over ten years and that our records showed that the current water flows were the highest they had been for a decade. The Bubbler and other springs of the Wabma Kadarbu National Park were outside the region where WMC had affected the Great Artesian Basin. There was no way that the Bubbler used to spurt water three feet into the air. The sand-saturated water bubbles no longer looked as spectacular as they used to because cattle and vehicles had been fenced out, and the sedge around the spring had grown and partially dammed the pool. The raised water level thus slightly reduced the size of the bubbles on the surface. If he had looked at the flow of water over the little waterfall on the side of the mound he would have seen that the flow had actually increased.

By this stage a group of the protesters had congregated around the car. The quasi-hydrologist authoritatively pronounced for all to hear, 'He's shittin' us.'

'Excuse me,' I interrupted, 'but who's shittin' who?'

I was determined that the activists should at least be aware of some of the facts. 'If you want I can come back and bring some reports with measurements of spring flows,' I offered. A happy smiling girl with a distracting tattoo mark on her forehead enthusiastically replied that they would appreciate some info. That was what I was doing the night that I met Mad Mark and was trying to strike up a conversation with Johnny Rambo.

'I don't know why you bother talking to those stinking, dole-bludging ferals,' I was told back at Roxby. It is easy to disregard the stance of all anti-nuclear activists as drug-warped dribble. They look, smell and act different. 'Ferals' are conveniently dismissed by mainstream society as destructive, parasitic, vandalistic wastes of protein. However, most are passionate about their causes, some are intelligent and nearly all of them have intriguing theories to extol. Everyone wearing a broad-brimmed Akubra is not a red-necked bushie and not all hard-hat wearers are small-minded miners. In the same way not all dreadlocked beanie wearers are drug-wasted ferals who are irrelevant to society. Without their extreme views and actions it is unlikely that the environmental ideals now embraced by most Australians would be as mainstream. I took every chance I had to discuss their issues.

The group that emerged from their tents, tepees, vans and ramshackle humpies near Lake Eyre South were a mixed bunch. There were old hippies, fresh-faced teenagers, angry disassociated youths and hard core activists as well as the spaced-out trippers like Mad Mark. The conversation around the fire switched from intrepid stories of blockades, arrests and subsequent court cases to elaborate tales about sinister conspiracies and cover-ups. Outrageous claims of the impact of uranium contamination clearly topped the bill. Fishermen supposedly caught a 2.5-metre fluorescent crayfish off the coast of Sellafield—a dirty old nuclear reactor and reprocessing plant in England. This nuclear industry was serious shit and it was their mission to stop it.

An unpierced teenager was sitting at the periphery of the assembled group, quietly sipping on her mug of gruel. Anomalies intrigue me so I moved over to her and started talking. I asked her why she had joined the protest camp.

'It's because your company is putting radiation into the artesian basin, into the water that the Aboriginals and wildlife drink,' she replied, confused as to why I would enquire about her motives when her evidence was so apparent.

'Where is this radiation coming from?' I asked in the most unchallenging tone I could muster. Her expression suggested that I had asked her where rain comes from.

'From the leaking tailings dam,' I was told.

I asked her if she was aware that the tailings dam was 100 kilometres from the Great Artesian Basin aquifers. I also said that the tailings that seeped from the dam only made it as far as the mine where they were pumped to the surface again. She looked like I had just informed her that rain came from rocks. I wasn't the only one to notice her unease. 'He's shittin' you,' said an angry-looking professional.

Since I had been labelled as a spokesperson for all things nuclear, a position that I had never volunteered, I thought I might as well play along. After all, it might help me to develop my own ideas about the pros and cons of the nuclear industry. I asked the group whether there was any realistic or even hypothetical developments that would coerce them to embrace nuclear power.

A budding physicist explained that plutonium—the most hazardous waste product of the nuclear fuel cycle—can now be rendered benign. However this technique was apparently not used to sanitise nuclear wastes and weapons stockpiles due to a conspiracy of the world's chief nuclear powers.

'Assume that this "conspiracy" was exposed and the new technique was then used to sanitise all of the nuclear wastes, would any of you then support nuclear power over fossil-fuel powered electricity?' I probed.

'No way, there is still the transportation issue. The shit is too dangerous to transport to the place where it would be sanitised,' retorted the spokesperson.

'Is radioactive waste any more hazardous or difficult to transport, or harder to clean up than ammonia, cyanide, chlorine, LPG or other hazardous chemicals that we truck around every day?' I asked, revelling in playing the devil's advocate.

'They should ban all that shit as well,' someone else offered.

Unlike Mad Mark, Johnny Rambo and a selection of more radical activists who were tired of my questioning, the girl with the tattoo on her forehead wanted to discuss environmental issues further. Her name was Izzy and together with her ex-geologist boyfriend Marc, who had worked in several Western Australian mines, they were keen to read the information that I had brought. I was impressed to see a couple who were obviously intelligent and committed to their cause, yet open-minded enough to discuss some of the issues that their campaign was targeting.

A few months later while sharing some roadkill snake that one of the omnivorous members of their camp had cooked, Marc, Izzy and I discussed our respective histories. I was keen to understand what made them tick and what their underlying concerns were. Both had been protesting, or 'active' as they put it, since the Jabiluka protests a few years earlier. Izzy's old man worked for ASIO, the Australian Security Intelligence Organisation. Although I was correct in assuming that he did not fund their activism, the source of Izzy's financial support was even more bizarre than I could have imagined. Rather than being on the dole like many of the activists, this bright, energetic and seemingly robust girl was on a medical pension. Izzy explained that she had been diagnosed as a 'manic elative'.

'A bloody what?' I exclaimed. Izzy proudly repeated that she was a manic elative. Apparently being permanently and unconditionally happy is a recognised psychological disorder. Wander in to a series of quacks and shrinks with a beaming smile and an infectious chuckle and you just might emerge on a pension. I wondered if her medical advice was to eat lentils in a showerless camp on a hot dusty hill near Lake Eyre for months on end until she had wiped that smile off her face. If so, the treatment had not worked.

Izzy and Marc were full of energy and ideas and were convinced that they were doing the right thing. Through a series of home video documentaries and raves they enlightened others to their cause. A collection of photos, newspaper and magazine clippings was plastered over the interior of Marc and Izzy's van depicting some of their 'conquests'. Izzy proudly showed me a photo of police rushing towards her after she had cream-pied Chuck Foldenauer, the Vice President of Heathgate Resources and project manager of the Beverley uranium mine. Like Roxby, Beverley was now a regular stopover on the activists' northern trail.

'Why did you cream-pie the main man?' I asked, trying to rationalise the value of stirring up the police and security guards by trashing a bigwig.

'There is a worldwide movement committed to cream-pieing prominent businessmen,' I was told. 'Bill Gates had been done, so have the directors of several other multinational corporations.'

'So it's sort of a notch in an activist's belt to pie a senior scalp is it?'

'No,' I was corrected. 'A pie is a great equaliser. We do it to show that no matter how much power and money someone has, they are no different from anyone else.' The activists saw it as their role to bring the tall corporate poppies back to earth and make sure that they were accountable for their actions. I was not convinced on their style but I couldn't argue with their sentiment about wanting their targets to be responsible for all the ramifications of their businesses.

Generalisations are difficult and can be misleading with diverse groups like the anti-nuclear activists, but I will take the risk. With the exception of the 'lifestyle activists' who are more interested in the actions rather than the issues, my experience is that most of them, like me, have a deep-seated abhorrence of nuclear warfare and the associated arms race.

Many of the activists have taken this stance further to conclude that anything to do with radiation, such as nuclear power or the disposal of used X-ray materials, is also bad by implication. These diverse issues are combined into an overarching evil symbolised by a radiation sign. Any debate that can attack these issues is set upon by the activists with vigour: workers' safety, mound spring contamination, Great Artesian Basin pressures, Aboriginal land rights, transport routes and environmental impacts all join the targets for protests and claims. It is hardly surprising that Mad Mark was shitting himself about meteorites.

Although some of the activists' concerns and actions were more extreme than most people's, it is fair to say that many of us continue to harbour concerns about the effects of radiation. Radioactivity and nuclear technology were thrust onto the world stage in association with a ruthless military device. Following World War II the world was held hostage for nearly half a century by the threat of a nuclear holocaust. As far as first impressions go, radiation got off to the worst possible start. I can remember at school when our class unanimously voted nuclear warfare as our greatest fear for the future. Anything nuclear was automatically conceived to be big, dangerous and out of control. Anything radioactive or using a nuclear energy source is still typically held in the same regard.

If it were not for the bombing of Hiroshima and Nagasaki, nuclear energy could well be viewed in a very different manner. The waters became even murkier following the weapons tests at Maralinga and the inexcusable recent atrocities at Mururoa Atoll, and other contemporary 'atom bomb' test sites. Chernobyl remains the most tangible link between the modern nuclear industry and a large scale environmental and safety disaster. This antiquated Russian power plant with pathetic safety systems has got as much relevance to the operations of the modern nuclear industry as chamber pots have to the management of modern sewer systems. Indeed the pro-nuclear camp present a strong case that properly managed nuclear electricity generation has proven to be both cleaner and safer than conventional fossil-fuel facilities. But to a society burnt by the Cold War, Chernobyl represented proof that anything using nuclear energy or involving radiation is an evil that must be stopped.

Like everyone involved in any industry that may have local, regional or global impacts I had to draw my own 'line in the sand' for the responsibilities I would bear. My 'line' was regional and my responsibility was the environment. But not everyone drew my line at the same place that I did. It is an interesting phenomenon to be considered a greenie at Roxby, and a apologist for the nuclear industry when I was among protesters or when I travelled to the big smoke.

In Armidale I was studying part-time at the university town that is at the same latitude but on a different planet to Roxby. At the great meeting place for students—the uni bar—I met a girl who was enrolled in Peace Studies. I was intrigued about her course and we chatted for a while about our respective studies. When I replied to her question about where I was from, she froze. 'That's the uranium mine isn't it? How can you dare to live in such a place?' she enquired. My new acquaintance was unreservedly paranoid about radiation.

Although I did not fancy becoming embroiled in a debate about the pros and cons of uranium mining I had to stick up for my home town. Although my words were falling on deaf ears I explained that in all probability Armidale had higher levels of background radiation than Roxby. Our radiation levels were lower than worldwide averages thanks to the thick layers of benign rock, clay and sand above the ore. Other locations, especially those with granitic bedrock close to the surface, experience double or triple these levels naturally. My disbelieving listener could not accept that the smoke that hung in the Dumaresq Creek valley in the mornings in Armidale indicated that elevated radiation levels were often trapped in the valley. When I suggested that she should visit Roxby to give herself a break from high radiation exposure she called her friends over who confirmed her belief that I was a whacko.

This was not an isolated incident. I got tired of being labelled a radiation-affected uranium miner. The last thing I wanted to do on a Friday night after spending a week trying to get my head around multivariate statistics was to get locked into winless, circular arguments. The day after my disembowelling by the Peace Studies contingent, I picked up some business cards from a landscape architect. For the remainder of my three-week visit to Armidale I told strangers in the pubs that my name was Andy and that I moved big rocks and bark chips around. I was human again and could participate in non-venomous conversations about music, sport and barmaids.

Even though I may have drawn my line around the local region, over a decade of continual bombardment, suspicion and sometimes outright horror or disgust made me look over that line. Was I correct to drive past those BP service stations all those years ago? I wanted to know whether this nuclear issue was really the world cancer that it is made out to be. I admitted while sitting in that uni bar that I didn't know the answer. Should we be mining the stuff at all? How and why did we get involved in the whole divisive and messy business in the first place? What was the history behind our involvement with radiation and the uranium mining industry? To me, as a South Australian, this naïvety was shameful.

❋ ❋ ❋

Laid out before the young geologist was an endless horizon that glistened white against the brown plains. The geologist had climbed Mount Painter in the northern Flinders Ranges to search for special rocks like the mineral he had recently discovered at nearby Radium Hill, and that he had named after his Professor, Edgeworth David. Both davidite and radium are radioactive, along with the uranium that he was studying at Mount Painter, and that was why he found them so interesting. The study of radioactive minerals was in its infancy but this geologist and others knew that these rare minerals contained properties that could make them extremely useful for researching the Earth's history. The year was 1910 and the young geologist was Douglas Mawson.

Mawson had been singled out as a promising young metallurgist by the great Madame Curie, who presented him with a gold leaf electroscope, an apparatus which at the time was state-of-the-art technology for detecting radioactive sources. Mawson went on to become one of the most intrepid, groundbreaking and decorated geologists and explorers that Australia has ever produced.

The value of radioactive minerals is that they emit energy and change their atomic structure as they decay. By investigating the atomic structure of the minerals, Mawson could establish how long they had been decaying, and hence how old the rocks were. The energy from this decay produced heat that in turn created sites like the nearby Paralana Hot Springs. Entrepreneurs opened these springs to the public in 1924 as a 'health spa', where the radioactive steam and waters apparently cured ailments ranging from gout to arthritis. Even as late as 1974 the South Australian Government still harboured plans to redevelop a tourist enterprise at Paralana. These are the same springs, a couple of hundred kilometres east of Roxby, where Mike Tyler discovered that nearly a third of the frogs were deformed due to the high radiation levels.

Although the radioactive steam of the Paralana Springs did not prove to be as healing as hoped, the energy produced by the radioactive decay was more useful. Another South Australian scientist who was also destined to become world renowned collaborated in the research that harnessed this energy. In 1940 the Germans had launched strikes against Norway and Denmark and were terrifying the western world with news that they were developing a super bomb. Unlike conventional bombs their 'atomic' bomb would unleash a massive explosion by splitting an atom of radioactive mineral. Mark Oliphant, the antipodean protégé of the great British physicist Ernest Rutherford, was coopted to help the British beat the Germans to their goal. Although neither the British nor Germans managed to develop the bomb during World War II, Oliphant was seconded by the American 'Manhattan Project' to assist them with the development of what turned out to be the first atomic bomb.

Oliphant's apprehension for these rapid developments was evident in a letter he wrote in early 1945. He recognised that the immediate result would be 'incalculable destruction' and was adamant that the use of such a weapon against an unwarned country was 'abhorrent and uncivilised'. He consoled himself in the knowledge that 'Though war has brought the opportunity to do these things … we know that in the ultimate analysis this aspect will be overshadowed by the benefits wrought for mankind.' The benefits to which he was referring were the provision of cheap clean electricity.

This gentle man, who was later known as a humanitarian and conservationist as well as a physicist, repeatedly campaigned against the use of atomic bombs both before and after the bomb was dropped on Hiroshima on August 6, 1945. I met him when he was in his 80s; his deep controlled voice and alert eyes broadcast a wisdom that could not be ignored. He retained both an anger and a guilt for the misappropriate use of the process that he had helped to develop in good faith. Not surprisingly, Sir Mark Oliphant was adamant that the raw materials of the nuclear technology should not be deviated to 'warlike uses'. He was also frustrated by the Cold War's negative influence on society's perception of nuclear technology. He wrote, 'People show anger with science for developing nuclear power. "How soon" they ask, "before this terrible force annihilates us?" They should instead be asking, how soon can it be made to serve us?'

Reg Sprigg is another reknowned South Australian and arguably the man in the best position to advise us on the benefits and hazards of the nuclear fuel industry. Reg was a student of Sir Douglas Mawson and he became internationally famous when he discovered and described the oldest animal fossils ever recorded; fossils found at a place called Ediacara, just across Lake Torrens from Roxby. He was a stalwart of Australian fossil-fuel discoveries and research, and his research was integral to finding the oil and gas basin now mined by SANTOS at Moomba.

Not only was he a geophysicist and palaeontologist, Reg was also a pioneering ecotourism operator and conservationist before such professions were recognised. He was an inaugural stalwart of the World Wide Fund for Nature and championed the Arkaroola–Mount Painter wildlife sanctuary, one of the first and most successful private wildlife sanctuaries in Australia. Reg's work was also integral to the discovery of uranium deposits in South Australia. For some of today's conservationists his enthusiasm for the uranium industry seems misplaced, yet Reg lamented the failure of society to take advantage of nuclear energy. He forecast a 'far greater catastrophe' than the dangers presented by the nuclear fuel cycle if we continued to exclusively use fossil fuels to generate electricity. This catastrophe would include the pollutants, acid rain and global warming attributed to burning fossil fuels.

If children at school were taught about the clean and efficient technology that the Swedes, Canadians and French were using to generate electricity they may have a far different view of nuclear fuel from that of their parents. Can you imagine dreadlocked protestors chaining themselves to dirty coal-fired power plants, pleading for more resources to be channelled into clean technology? Ironically, the original coal-fired power station at Port Augusta was named in honour of the former Premier of South Australia, Sir Thomas Playford, who was one of the most enthusiastic promoters of uranium exploration in the State. Can you imagine Joy Baluch, the passionate and outspoken Mayor of Port Augusta, campaigning vehemently for the Thomas Playford power station to be replaced with a cleaner nuclear plant? Can you imagine outback communities vying to safely store the wastes from power plants in synroc, deep within some of the most stable rocks on Earth?

Some eminent scientists reckon that the nuclear energy industry is inherently good given the present options. By the same token, nuclear power generation is still not as cheap or clean as Playford had hoped. We will be judged by future generations for the degree that we do or do not embrace nuclear energy. Hindsight has a habit of turning past heroes into villains and vice versa. We recall with horror the inhumanity of the 'stolen generation' of Aboriginal children who were taken from their families by an institution that believed it was operating in their best interests. Similarly, today we pay people to plant trees to control salinity and erosion on the same land that we paid people to clear a few decades ago.

Future generations may lament our wasteful and polluting dependence on fossil fuels. The thousands of litres of fuel that I used driving in search of taipans, plains rats and knobtails may, in retrospect, be viewed with the disdain that we treat the old bounty on Tasmanian tigers. Time will tell how the environmental consequences of the nuclear energy industry ultimately stack up against the greenhouse effect, battered ozone layers, desertification, forest clearing, pollution, overpopulation and consumerism which, like it or not, our society is currently embracing.

It is difficult to perceive how future generations will judge the pros and cons of nuclear energy when today's society cannot reach an informed consensus. One reason that many Australians, including myself, are unsure about the whole issue is because we are exposed to so much distortion of the facts. Pro- and anti-nuclear factions remain embroiled in the divisive Cold War of 'green-washing' versus 'brown-washing'.

'Green-washing' is the trumped-up or exaggerated claim by industry or government that environmental issues are being addressed proactively and satisfactorily, when perhaps this isn't the case. From an insider's perspective I should be in a good position to identify green-washing. To my knowledge, apart from some naïve rather than insidious comments by over-patriotic mine supporters, WMC's reports and authorised statements have been clear of this washing. However, I distinctly remember when the first annual report I wrote was returned by a senior manager. Such was his concern about the report being distorted out of context that I was instructed to remove all reference to the words 'kill', 'shoot', 'trap', 'succumb' or 'death'. 'Mining companies do not kill anything,' I was told. I was perplexed and frustrated as to how I would describe the methods or outcomes of the extensive feral animal 'control' conducted on the lease. Thankfully, paranoia about being open and honest with reporting impacts has now been largely overcome, but I did not expect sceptics like Marc and Izzy to believe our reports on face value. That is why many companies now seek auditing of their environmental reports from respected third parties, so that they can confirm that their figures have not been green-washed.

Marc and Izzy's predecessors campaigned passionately to view the once restricted environmental monitoring results from Olympic Dam. Ever since 1991 these reports have been freely available. The response to these public reports which include innovations and new discoveries has been deafeningly quiet. Rather than taking an active interest and potential role in environmental management, most activists are preoccupied with the opposite of green-washing. Brown-washing is not just pulling out the dirty laundry for all to see, it usually involves chucking mud on the clean stuff as well.

The mud includes headlines or graffiti like: 'Town pool closed due to deadly layer of Radon gas', 'Olympic Dam is an ecological disaster area', 'Roxby rainwater contaminated' and 'Don't pump the Lake'. Unfortunately, with brown-washing blurring reality and perceived global threats misrepresented by misconceived local scenarios, the issues associated with Olympic Dam and indeed the whole nuclear industry are too diffuse to pin down. Consider the safety figures alone. Widely circulated 'green' newsletters claim that 250,000 people have died as a result of the Chernobyl tragedy, whereas the United Nations Scientific Committee on the Effects of Atomic Radiation has recently confirmed that fewer than forty people died, and that approximately 2,000 people contracted radiation sickness or leukaemia as a result of the accident.

Who can you believe? Who's shittin' who? How can we decide on the relative risks or advantages of different methods of generating electricity, mining or other contentious issues when the information we receive has been grossly manipulated by those with preconceived ideas?

What then without any 'washing' is the impact of the Olympic Dam mine on the local and global environment? Catch up with a crested bellbird or beaked gecko on the red dunes within a couple of hundred metres of the smelter and they will probably cough out that they are not too impressed. Ask those same species a couple of kilometres from the smelter and they would probably deem that the mine has had negligible effects. Visit an activist camp, listen to some of the callers on talkback radio or have a drink in the bar at an eastern state's uni and the answer will inevitably be different.

When my grandchildren are passing their judgement on my involvement at Olympic Dam I hope that they will still have the opportunity to meet with the likes of Izzy and Marc, and also with inspired scientists like Mawson, Sprigg and Oliphant. If they can be challenged by different perspectives without hacksaws, spray paint and cream pies, or the more insidious green- or brown-washing, society will be better able to determine the best paths to take.

20
Soft lighting

One evening at the annual conference of the Ecological Society of Australia I was chatting with Hugh Possingham over a quiet ale. Hugh is one of Australia's leading ecologists and he achieved the ranks of professor when he was scarcely out of his teens. I ribbed Hugh that despite his high profile and blossoming academic career, I had not seen him or his students presenting any papers at the conference. To my embarrassment Hugh replied that he was chairing a symposium on metapopulation dynamics the next day, during which many of his students were presenting their research. 'That sounds interesting,' I said. I lied.

Hugh combines his passion for conservation and ornithology with the applied mathematics that is his forte. I had overlooked his upcoming symposium on metapopulation dynamics because I had no idea what it was about and a first glance of the titles indicated that Hugh's students were talking about mathematical models. I could think of nothing more boring than listening to a series of nerds who were too 'intelligent' to study real animals, spruiking on about mathematical equations that were intended to represent hypothetical species.

The following day I found myself trapped in Hugh's symposium. For the first half an hour I ignored the equations and mathematical jargon while lamenting that these students were wasting their time with irrelevant hypothetical postulating. Why weren't they studying ecology and helping to secure the future of threatened animals such as the inland taipans, knobtails, and lake-dwelling dragons that I was concerned about? Then I heard one of the students mention 'probability of extinction'. I tuned in instantly. This bloke was going to predict the likelihood that a particular species would go extinct. Although he was still talking about species A and populations X, Y and Z, I figured that he might provide some clues to the conservation of real animals. Hugh Possingham obviously thought so too!

I learnt that 'metapopulation' referred to a species that relied upon a number of populations for their long-term survival. Hugh's students went to great lengths to demonstrate that the survival of these metapopulations depended upon the ability for new populations to be formed as frequently as others went extinct. Areas of unoccupied suitable habitat were nearly as important as areas where animals were currently living. I switched off from the seminar as I considered the implications of these models for my rare species around Roxby.

The Pernatty knobtails in the dunes near Nurrungar are a separate population from those at Lake McFarlane and Lake Windabout. Very occasionally the gene pools of these different populations mix when individuals breach the rocky barrier between them. According to the theories I had just heard, the long-term survival of Pernatty knobtails was largely dependent upon maintaining as many of the discrete populations as possible. A species has a much greater chance of surviving in a changing world if it is made up of several populations.

If, for example, a fire caused the extinction of the Nurrungar dunes mob of Pernatty knobtails, these unoccupied dunes would probably be recolonised by another nearby population. Ten, or a hundred, or a thousand generations later, the establishment of a golf course could wipe out the Lake Windabout population. But when locusts destroy the irrigated fairways and the sand dunes reclaim their rightful place, knobtails from Nurrungar may in turn recolonise the Lake Windabout dunes.

The ill-fated dodo, which was hunted to extinction on Mauritius by whalers and feral pigs in the 1600s, did not have the same ability to ward off extinction as do Pernatty knobtails. The dodo's biggest problem, apart from being slow, dopey and tasty, was that its entire species was made up of one population. The whalers and their pigs wiped out the Mauritius population and the whole species was gone. There were no second chances.

As the lecture continued my thoughts turned to the Lake Eyre dragons. These unique little lizards are found in three populations with the vast majority occupying the lake from which they take their name. They also occur on lakes Torrens and Callabonna. Unlike their prime habitat on Lake Eyre where they shelter under the thick salt crust, Lake Torrens does not have a salt crust. I have slipped around on this mudflat of a lake in several unsuccessful searches for dragons. Although they are apparently not common, the Lake Torrens population may be exceptionally important for the survival of the species. If any future climate change renders Lake Eyre unsuitable for these dragons the species may depend on colonising other local salt lakes where the crusts are thin or lacking altogether. Individuals adapted to conditions more similar to these 'new' lakes will be better positioned, both geographically and ecologically, to replace Lake Eyre as an important habitat for the species.

Hugh's students helped me to realise the serious implications of the 'greenhouse' changes in temperature and rainfall for our rare plants and wildlife. More recently I learnt of an even more insidious effect of the predicted doubling of atmospheric carbon dioxide levels within the next century. One of the first biology lessons we all learn at school, even before the one on Von Frisch's bees, is that growing plants convert carbon dioxide to oxygen through photosynthesis. Increased carbon dioxide should therefore induce greater plant growth and provide more food for herbivores, correct? Wrong! Plant tissue actually contains less protein when grown in a carbon dioxide rich environment, which in turn can result in the slower development and a higher mortality in herbivores. Changes in plant growth rates and nutrient concentrations will affect every animal on the planet, including us. Arrggh!

It is difficult to conceive that possibly the greatest threat to some rare outback inhabitants and other life on Earth is carbon dioxide. If gulls were the only problem we could bait them. Bulldozers, chainsaws or radiation, if they were the only cause, could be stopped by fair means or foul. However, when the real enemy is increased levels of an atmospheric gas I feel more out of my depth than trying to comprehend the convoluted equations of metapopulations.

It is easy living in the sparsely populated outback to consider greenhouse gas emissions to be a problem of the big cities and industrial regions. After all it is in these areas of high population density where the bulk of these air pollutants are generated. For confirmation look at the smog layer as you approach any of our large cities or power stations on a still day. It therefore seems rational that efforts to address climate change should focus on these high-polluting regions. If we are going to develop and test 'green' non-polluting power generating initiatives, it superficially seems most appropriate to concentrate efforts in the cities. However, the reverse is true.

Our large cities have relatively cheap supplies of mass-generated power for households, businesses and transport. By contrast, inhabitants of remote localities are faced with considerably higher power costs because we typically have to generate our own power. Furthermore, due to searing summer temperatures and the need to refrigerate large stores of food, our average demands for power are usually higher than those of most city dwellers.

With the exception of a few ingenious turbines that have harnessed energy from flowing artesian bores, diesel generators power most outback homesteads and towns. These generators are not only dirty, noisy and expensive to run but they also contribute to those greenhouse gases in the same way as domestic power plants. Due to the huge expense of transporting fuel to our remote locations and maintaining diesel generators, the economics of green power stack up more favourably in the outback than at places connected to the electricity grid. Because of these factors outback residents are best placed to make economic and environmental savings by adopting renewable power initiatives. The widespread efficient use of the sun or wind to generate electricity may eventually stem from trials conducted in these remote locations.

Izzy and Marc thought that we should all be using renewable energy for everything. Marc was enthused about a renewable fuel technology called an 'orgone accumulator'. This device apparently used water as a fuel source. With two glasses of water that had been energised for half an hour or so you could power a light globe for three years. With the correct design of concentric stainless steel funnels and fibreglass rods, electricity could be generated from water by harnessing the immense forces of the Earth's energy fields. I was not able to ascertain whether these fields were gravitational or magnetic. My interpretation of Marc's understanding was that it was a hippy type of energy that was difficult to describe to a sceptical scientist. I used the word 'sceptical' as often as they used 'conspiracy'. If it was that easy to generate electricity why didn't they have a water generator with them? Marc had just received the instructions and said he was going to try to build one. I believed him.

What I liked about Marc and Izzy was their lack of hypocrisy. They were concerned for the environment and revelled in demonstrating innovative ways to generate energy without burning fossil fuels. A wind-powered generator mounted on their van powered their mobile 'cinema' and scant lighting at the camp. Because I was sceptical they showed me their modified fuel tanks, complete with a radiator heating system, that enabled them to run their van on waste food oils. Sometimes their white exhaust smelt of fish and chips, sometimes doughnuts. I admired anyone who was prepared to collect buckets of waste oil from the fast food outlets that they passed on their travels, and who was prepared to stop every 100 kilometres or so to flush floaties from their modified fuel filter.

The whole activist camp was an impressive example of frugality compared to our shockingly wasteful society. The activists each spent only about $30–$40 per week on food and wasted very little. Their electricity costs and rubbish-collecting levies were non-existent. But these were no 'ordinary' people and they were not planning to live like this forever. It is no coincidence that the camp was occupied during the mildest months of the year. Several local couples demonstrated that a lifestyle without mains power or chugging diesel generators does not necessarily mean that we have to revert to the 'primitive' existence of an activists' camp.

Greg Emmet and Prue Coulls from the Coward Springs' campground near Lake Eyre, live for much of the year with a couple of solar panels and batteries as their sole power source. Although they don't have microwaves or electric hot-water systems they enjoy a comfortable standard of living. For an investment of less than $4,000 Greg and Prue generate 240-volt power that allows simple lighting, phones and Internet access. Although our pioneers would have revelled in such luxurious conditions there are few people today who are prepared to tough it out in summer without an airconditioner, or at least a big fridge and a storm of fans.

Even more isolated than Coward Springs are some of the far-flung homelands in the Anangu-Pitjantjatjara (A-P) Lands, in the far north-west of South Australia. One would expect that the drone of a diesel generator would be one price that the Aboriginal inhabitants of these homelands would have to pay for living in some of the most spectacular wilderness in Australia. But the only sounds competing with the yapping of dogs and the excited squealing of children are the typical domestic sounds of a TV, washing machine or airconditioner. Banks of solar panels on the rooves only leave room for the solar hot-water services. By generating sufficient power to run domestic appliances twenty-four hours a day these houses point to future energy-generating prospects for other remote homesteads.

Already most homesteads in the outback are installing inverter systems that charge batteries while their generator operates for a few hours each day. Then when enjoying the serenity once their generator shuts off, their batteries continue to power lights, refrigeration and the Internet that they can use to research other power-saving initiatives. With increasingly popular hybrid generating systems that may incorporate solar, wind or hydro power the energy savings are even greater. Outback dwellers are more likely to experiment with these innovations than are their city counterparts because they are becoming increasingly familiar with the technology. Solar pumps are replacing outdated and dangerous windmills and diesel pumps, and solar-powered electric fences are also becoming more common, with even the Dog Fence converted to solar power in some areas. For generations outback residents have monitored and limited their use of water and fuel, which their suburban cousins take for granted.

Remote homesteads are not the only houses that are now benefiting from energy saving innovations. Mick and Trish Evans' purpose-built 'environmental house' at Roxby has been orientated with the living areas on the north and bedrooms on the south to minimise both heating and cooling requirements throughout the year. Colonists of outback Australia quickly recognised the benefits of the wide verandahs that are now characteristic of most homesteads. The Evanses have taken this icon one step further with a simple yet ingenious development. Their slatted verandah on the north side allows the low angles of the winter sun to penetrate but shades the walls and windows from the scorching heat of the summer sun. High ceilings equipped with fans and woollen insulation also keep the rooms cool in summer while the rest of the town swelters. Along with their solar hot-water system, which is mandatory on all recently built WMC homes at Roxby, they have also installed solar panels that provide eight to twelve kilowatts of power a day, depending upon the amount of sunshine and the temperature. Because they only use appliances adorned with a constellation of stars on their door the Evanses use a meagre seven kilowatts of power a day in winter. As a result they sometimes feed power back into the electricity grid, which provides enough credits to cover their peak usage of up to thirty kilowatts in the height of summer.

It is not only family homes like the Evans' and those in the A-P Lands that can benefit from renewable energy. An impressive-looking 150-kilowatt wind generator dominates the skyline of Coober Pedy. Despite its early promise this trial generator has not proven to be worth the investment. The giant windmill only saves the council approximately $60,000 worth of diesel from its annual expenditure of over $2,000,000. Neville Mitchell, Coober Pedy Council's electrical engineer, laments that 'his' wind generator built in 1991 has not been able to cope with the harsh local conditions. Apparently the huge white windmill has consumed considerable time and effort and generated more angst than electricity. 'We've had a go with this one, let another town put their hand up for the next one,' Neville sighed when I asked whether he would consider trialling an updated wind generator.

Despite the teething problems and inefficiencies experienced at Coober Pedy, other outback communities are experimenting with renewable energy supplies. With soaring diesel costs and ever-increasing concerns over carbon dioxide pollution, wind and solar generators are becoming increasingly attractive. Plans are being developed for the world's largest commercial wind farm near Elliston on the Eyre Peninsula. On the other side of the peninsula the Whyalla Council are also investigating the feasibility of wind generators at nearby Point Lowly, along with an innovative 'solar oasis' generating both electricity and desalinated water. Power requirements in the houses and offices at Umuwa, the administrative 'Canberra' of the A-P Lands, are reduced by installing sunlights so that lighting is not needed during the day, and by equipping buildings with solar-water heaters and low-wattage lights. A 'sunfarm' planned for Umuwa will provide 200 megawatts of power from ten huge solar collectors to complement the diesel generators that supply power to most of the regional settlements. The white elephant at Coober Pedy will hopefully be seen as a useful evolutionary step in the development of viable renewable energy generation for remote localities.

'Green' power generation that is economically attractive in the outback may significantly contribute to domestic power requirements in the cities. Already consumers are able to buy a certain percentage of 'green' power at a premium. If more city dwellers adopted the energy-saving innovations like the increasing number of outback residents, Australians would be in a far better position to proudly claim that we have accepted our responsibilities to address the greenhouse effect. On the other hand, while society continues to rely on the easy option of burning fossil fuels for bulk power generation and transport, we continue to shake the climatic tightrope for many of our plant and animals species in the outback and beyond.

21

Treading lightly

The Dalhousie ruins shimmered in the distance, distorted by the heat haze that rippled like a lake above the stony plain. The roads had just been reopened following a heavy rain so the dusty convoy of cars that typically converge towards this tourism mecca was missing. Together with Fewster, Matthew and Mick, all of whom had survived Lake Eyre walks, I had successfully crossed the flowing Irrapowadna Creek near the South Australian–Northern Territory border. Just when we thought we had negotiated our last obstacle on the way to the Dalhousie Springs we became hopelessly stuck in the next nameless creek. The Landcruiser's wheels spun downwards in the liquid mud as we tried in vain to drive forward or backward out of the mess. Like every other occasion when we got bogged we first tried clearing the mounds of mud in front of and behind the wheels. But once we gingerly lowered ourselves to our knees we could see that the mud had already moulded around the dif, sump and springs. The glue-like clay clung to our shovels like a frightened koala to a gum tree, so we had to lie in the mud and scoop it out from under the car with our arms.

Every armful of mud got heavier, stickier and less rewarding as it oozed back into the low depressions that we excavated around the wheels and axles. After we had eventually exposed the top two thirds of the tyres we spread out across the barren plain to find anything larger than small round stones to use for traction under the tyres.

Under a branch that had washed down the river I found a tiny Grey's skink—a delicate little lizard with four fingers and five toes. I dutifully measured it and recorded its location and then set to work in the mud again. About twenty minutes later the others came back with a couple of sheets of iron that were littering the gibber plain near the Dalhousie ruin. While I had found Australia's smallest lizard they had seen Australia's largest—a perentie—at the ruins. The following day we measured and photographed this enormous aloof reptile and returned him to his castle near the date palms.

The extremes in lizard sizes complemented the landscape and climate. We were bogged in a muddy flowing stream on a scorching hot day in the desert. But the biggest extreme was yet to come. Our spirits soared after we eventually extricated ourselves from the slop amidst spinning tyres, screeching iron and a burning clutch. Fifteen minutes later, absolutely covered in mud and sweat, we drove up to the main Dalhousie spring. It was January 1990 and the first time any of us had been to one of the outback's most spectacular attractions.

The peanut-shaped tree-rimmed pool beckoned enticingly. I jumped from behind the steering wheel, threw off my clothes and charged at the pool. It was clear and deep enough to risk one of those nervous flat dives that you do when you are seduced into water before you can think.

Whack!

I did not know what had hit me. My world blackened. After a split second I realised that nothing had hit me, but that I felt like a yabby that had just been thrown into a boiling pot! I recalled the stories of dogs that had boiled to death after jumping into boredrains on the Birdsville Track. I surfaced like a skimming rock, lunging out of the water and taking a huge breath. I had been so intent on cooling off in the clear spring that I had forgotten that the water was actually about bath temperature.

Once I recovered from the shock of dealing with what was clearly not a near-death experience, I surveyed my surroundings. A dense fringe of white-trunked tea-trees clung to the edge of the spring pool and lined the gushing outflow creek for as far as I could see. Dragonflies and damselflies created perfect reflections in the mirror of the water. Clean white-breasted woodswallows and iridescent rainbow bee-eaters perched attentively on dead branches as they searched for insects. A pair of black ducks that we had disturbed made a noisy departure. This surely was an idyllic climax to our travels.

I looked down into the water to try to locate a Dalhousie catfish, goby, or hardyhead (all unique to the Dalhousie Springs), but not only were the black ducks disturbed by my impatient entry. Although my dive had not resulted in any personal injury it had damaged the very place that I had driven to experience and enjoy. A mushroom cloud of silt was bellowing from where my feet had pushed off the floor of the spring. The mud that had caked my body minutes earlier was also dissolving off and clouding the primeval spring water.

Looking around I could see the damage caused by earlier visitors. Branches had been broken off some of the tea-trees for firewood. Other trees had been burned, presumably by a camper's fire that flared out of control. The barren salty apron near the pool where our muddy Toyota stood had been the car park for thousands of cars over the past fifty years. Each vehicle had contributed to the compaction of the soil and the destruction of the vegetation. As a result run-off washed into the spring, contaminating it with clay and salt. No doubt many visitors had unwittingly, or even deliberately, introduced soap, detergents or cosmetics into the unique environment that they had travelled so far to experience. On a later visit I noticed little rainbow-coloured oil slicks emanating from pale travellers who had lathered themselves with suncream and insect repellent seconds before entering the water. Broken bottles and strands of soiled toilet paper indicated that some of these visitors did not deserve to enjoy such a biologically and culturally significant place.

Dalhousie, and indeed the entire outback, is now a real drawcard for travellers. The Dalhousie Springs are rarely unoccupied and during the peak winter months it is not uncommon to have thirty or more 4WDs jostling for the best campsites. Such a large number of people cannot enjoy such a location without ruining it unless they are meticulously managed. As much as I hate to admit it, my first visit was ample proof that without proactive management, places like Dalhousie would inadvertently be loved to death.

Most of us yearn to holiday at isolated pieces of paradise that yield no scars of previous visits and provide no bins, designated fireplaces, car parks or rules that remind us of the world that we have left behind. Unfortunately such dream locations are scarred by each and every visit no matter how careful we are. Fortunately the National Parks and the Irrwanyere traditional owners now jointly manage development works at Dalhousie. Their initiatives aim to preserve the key features of these spectacular springs and the surrounding country-side, while enabling travellers to still experience one of the outback's most fantastic sites. Vegetated earth bunds now prevent salty polluted water from rushing into the main spring. Fencing prevents vehicles from being driven into sensitive areas, and toilets, showers, signage and fireplaces (with provided firewood) are all designed to minimise the impact of visitors.

For some people these developments detract from their bush experience, but they can have numerous unexpected benefits. Information displays allow naïve travellers to better appreciate the biological and cultural significance of these areas. Satellite phones provide an invaluable safety net in case of an emergency. Visitor books accumulate the musings and suggestions of thousands of people. Not only can tourist facilities minimise unintentional environmental damage, they can demonstrably improve the environment as well. In 1991 Greg Emmett, a carpenter-come-stonemason from Kangaroo Island, started to clean up the rubbish-ridden and dilapidated Coward Springs' railway siding near Lake Eyre South. From his solar-powered base he converted the mess into a rustic but comfortable campground and fenced off the boredrain from cattle, resulting in an increase in the number of birds attracted to the wetland.

For several years I had slopped and slipped around the sedge and mud at the edge of the boredrain to compare the bird usage of the Coward boredrain with the nearby mound springs. In summer this wetland bird count was inevitably followed by a refreshing dip in the borehead that was accessed along a steel rack pushed out through the reeds. Later Greg ingeniously built a spa from old railway sleepers and provided the perfect mud-free cooling-down spot for many hot dusty travellers. This spa has the added attraction of being one of the few places where you can feed wild lizards. Wall skinks live in the cottages and campground toilets and quickly colonised the sleepers that form the screen around the spa. These little grey skinks have an amazing ability to cling to the vertical walls and are always alert for flies that have been attracted to visitors. After a while they will even get plucky enough to scurry along your legs and leap on live flies, to the delight of anyone patient enough to remain motionless.

Just after Christmas in 1992 I was doing my normal walk through the Coward boredrain and I flushed an owl that was hiding in the low sedge near my feet. I had occasionally recorded a barn owl roosting in the nearby date palm and I assumed that it must have now been sheltering near the water to avoid the heat. I had only taken a few more steps when I flushed another, then another, then yet another owl from the low sedge. They all flew a short distance then dropped back into the sedge again. This was strange indeed. I remembered that there was another type of owl from Queensland that looked like a barn owl but roosted on the ground. Not knowing what the Queensland owl looked like I flushed these birds again and took some notes.

They were not as white as barn owls, and since I am not very good with the names of subtle hues I opted to describe them as 'rufous-fawny brown'. I learned later that their breasts are in fact a 'washed orange' according to one field guide, or 'mostly cream with a strong buff wash' according to another. Bec Mussared, a friend who was with me, noticed their black wingtips and the black bands on their tails. I raced back to Roxby to look up the birds in a field guide.

Frank Badman told me that he had never seen eastern grass owls before but he knew they had been recorded only three times in South Australia, always in the far north-east of the State. In subsequent months I counted up to ten grass owls at Coward, and the list that is maintained by Greg's partner Prue attests to the continued value of this renovated site for both birds and campers. Even the most ornithologically challenged travellers delight in hearing the strange squawks and hoots of crakes and swamphens as they wash off the bulldust in this relaxing spa.

※ ※ ※

Outback tourism is already big and it is still growing. Already the tourism industry generates twice the profits of the pastoral industry and is second only to mining as a revenue earner from Australia's rangelands. Right throughout the region tourism needs to be managed because no matter how well intentioned most travellers are, they still have an impact. Most of us know of parks or places closer to towns that have been loved to death, spoiled because too many people have visited without appropriate management.

Some pastoralists and outback entrepreneurs now cater for and profit from this boom industry. Although rubbish, off-road driving and the over-exuberant collection of firewood causes environmental problems, the move to showcase the outback should reap environmental benefits. Travellers do not want to visit sites degraded by overstocking, feral animals, mining or neglect. If I asked the native animals whether they would rather live in areas set aside for low-key tourism rather than more invasive land uses, I reckon I could predict their answer.

Listen to international tourists recounting their favourite memories and they will rarely be of flash resorts or manicured campgrounds. Having their bags carried by concierges through the cold lobbies of mere five star hotels is hardly a postcard experience.

More likely they will rave about the sunrise, sunset or night sky from their isolated million-star campsite, or the dingo, goanna or snake that was seen from their tent. For many escapees from the rat race their most vivid experience is of simply being in the wild without any trace of the civilised world. Others enjoy interaction with Aboriginal people, pastoralists, miners, biologists, truckies, artists and the others who call the outback their home.

Reputable ecotourism outfitters can provide such experiences with minimal impacts to the environment. For over a decade Andrew Dwyer of Diamantina Tours has thrilled international and interstate travellers by setting up camps at out-of-the-way locations near Lake Eyre, giving visitors just the experiences they desire. Impressively these sites are used repeatedly and remain nearly as pristine as when Andrew first started using them. By being self-contained and emphasising the need for travellers to minimise their impacts, Diamantina Tours are a great example of sustainable outback tourism. Other tag-along-tours and indigenous tourism outfitters have also started showing their clients how to enjoy central Australia without leaving behind the legacy of chain-sawed trees, generators, broken bottles, toilet paper and off-road driving.

In the same way that Australia has benefited by embracing many different cultures, the outback environment is likely to be a better place for supporting many different land uses in an appropriate mix. Vast tracts of land allocated solely to pastoral, mining, indigenous or even conservation initiatives will inevitably suffer from a lack of awareness, funding, skills or willpower. The tourism industry has the responsibility and ability to inform visitors of both the wonders and the threats to the outback. Indeed, tourism may be one of the saviours, rather than one of the problems for the outback environment.

22
Cherish or perish

Concentric rings of sand surge upwards and outwards from the unseen depths of the pool perched on the summit of the low mound. Without warning a mighty belch suspends a pillow-sized bubble of water above the surface for a few seconds before it is sucked back down into the depths of the spring again. A constant stream of water cascades down the waterfall a couple of metres away. Despite what Johnny Rambo would have you believe, this broiling cauldron is behaving just like it has for thousands of years.

Aquatic isopods resembling slaters bulldoze through the spring, only stopping when the shadow of a heron or a human alerts them to danger. Eons ago, as today, you would have also seen the tiny poppy seed-like aquatic snails and their slightly larger, cone-shaped amphibious cousins clinging to partly submerged rocks. It is easy to imagine that the environment around the Bubbler is unchanged from when the enormous nearby Mt Hamilton spring was active many comets ago. But look a bit further, beyond the lush green sedge of the spring, and you will get a depressing shock.

A small boardwalk now snakes from the car park through pillars of eroded white sand that are capped with one of the most unpalatable of all desert plants—the nitre bush. Two-hundred years ago Aboriginal people sat cross-legged on ground that was a couple of metres higher than the boardwalk, chipping rocks to make tools or cooking wildlife that they had caught near the spring. The land and the vegetation that these people lived among looked very different to what it does today. For the Bubbler was the site of one of the most devastating environmental catastrophes that the outback has ever experienced.

In the 1860s Peter Ferguson took up the lease of the recently proclaimed Mt Hamilton Station, which incorporates much of the land that makes up today's Wabma Kadarbu Conservation Park. Mt Hamilton Station was blessed with permanent water from the Bubbler and other nearby springs. Following rains in the Stuart Range, the Margaret Creek flooded out across the plain on its way to Lake Eyre leaving behind a lush floodplain. Small depressions, or gilgais, on the stony plain trapped run-off and in turn supported dense patches of saltbush and grass. Ferguson had it pretty good, in fact he had it too good. The ample feed and endless supply of water enabled him to amass a huge flock of sheep and a herd of 4,000 cattle.

Only a few years later these high stock numbers combined with the lack of feed during the drought of the mid-1860s to devastating effect. Ferguson would have been better off fishing in the mirages that shimmered his horizons than maintaining livestock in the desert that he was creating day by day. The fledgling pastoralist watched helplessly as all of his sheep died and his cattle herd dwindled to only 200 pathetic survivors. Back then, Ferguson had no option but to grieve as his animals and livelihood bellowed their last mournful cries of anguish. By the time they had perished, the sheep and cattle had removed every palatable morsel within walking distance of the mound springs. Rain and wind scoured the exposed soil, destroying less palatable vegetation and ruining the ability of the country to regenerate. Ferguson could not have done more damage with 100 bulldozers.

What is tragic is that this disaster was not restricted to the Bubbler area or to Peter Ferguson. During the same drought John Chambers lost 4,250 cattle on the neighbouring Stuart's Creek Station, and John Weatherstone was left with only 245 cattle from his herd of over 3,000

Like its orange cousin, the pale green western brown snake at Roxby is only found in sandy country.

The large brown 'witjita' form of western brown snake characteristically flares its neck when threatened.

The slender orange western brown snake, with its characteristic black head, does not grow as large as its 'witjita' cousin.

The Goulds goanna will usually try to avoid battles, but will fight if necessary.

Two chestnut-breasted whitefaces, a rare bird species restricted to outback South Australia.

Is the extra arm on this trilling frog at Roxby caused by radiation or other pollutants?

After being dug into the hard clay, these 401 pitfall traps revealed an amazing diversity of reptiles and an abundance of frogs at Roxby.

An apparently healthy red-backed kingfisher chick looks out from the remains of its nest built into radioactive ore at Roxby.

Beautiful and rare grass owls occasionally make the renovated Coward Springs bore drain their home.

This fenceline shows that long-term overgrazing can lead to barrenness in the outback.

The author proudly holding a Woma python caught in the Cobbler Sandhills. (Photo: J Fewster)

The longest man-made structure in the outback protects rare hopping mice from cats.

Katherine scanning unsuccessfully for rock wallabies from our sleeping cave in the Davenport Ranges.

Extinction still occurs today. Katherine holds the remains of the last Davenport Ranges rock wallaby.

Katherine proudly declaring that after three years of frustrating work, the Arid Recovery Reserve is finally free of rabbits.

on nearby Finniss Springs Station. Some of the springs were probably buried under drifts of all-consuming white sand; even those that survived as pugged bogholes were too far from ungrazed vegetation to be of value to stock.

Much of this early damage was irreversible. Comparisons of both sides of old fencelines prove that some of the barren plains with which we associate much of the outback were once vegetated pastures. Removal of long-lived and sensitive vegetation, erosion of topsoil and the extinction of animal species have now changed the nature of the land forever.

Unfortunately pioneering pastoral disasters were not limited to the 1800s. Brown–grey photos of waterholes, railway sidings and homesteads show that livestock and vegetation seldom coexisted even in the good years of the early twentieth century. Many if not most of the early stations in both sheep and cattle country experienced tragic stock losses that were invariably preceded by the massive degradation of the land around permanent waters. Like Ferguson, Weatherstone and Chambers, most pioneering pastoralists had scant comprehension of the fickle nature of the outback climate, and even when they did there was little they could do to save their stock or their land when the droughts hit.

No wonder they had dust storms that blocked out the sun for days. The dust storms of today are mere 'dustpan' storms compared to the horrific dust storms of 100 years ago. The pioneers used to suffer 'wheelbarrow' storms where sand ripples invaded their hallways and verandas like an inland beach. On the Birdsville Track they even had 'bulldozer' storms where sand drifts buried entire buildings. These storms not only affected the pioneers' homesteads but they suffocated animals and buried or sandblasted much of the vegetation that had been spared by the rabbits and stock.

Mining and tourism sites in northern South Australia would not cover as much as a fingerprint on a map the size of this page. By contrast cattle, and to a lesser extent sheep, have access to an area the size a small handprint. As a result, modern-day pastoralists have inherited custody of most of the outback. This small resilient community has far more potential to either improve or destroy the environment than all of the conservation workers, miners, greenies and tourists put together. At the same time, the pastoralists probably have fewer resources than any of these groups. Using such vast tracts of land in a wise and sustainable manner is not only a huge responsibility but it is one of the most difficult jobs around.

Just as it is for the plants and animals of the outback, the unpredictable nature of the rainfall is the major challenge faced by pastoralists. Were rainfall predictable, accountants or even politicians could probably manage a pastoral station. However it is not that easy out here. Can you imagine an accountant developing a five-year financial plan when the difference between annual profits and losses is dependent upon a totally unpredictable factor? How would a concerned politician cope with not being able to predict which issues are most relevant to their electorate? Pastoralists have no control over the single most important factor that affects their business and their land.

I am scathing of long-range weather forecasters because of the potential damage that they can cause. Forecasts of impending rains are sometimes too tantalising for some pastoralists to ignore. Many bushies also like to think that they can predict big rains, regardless of what the planets, sunspots or ocean temperatures suggest. They ascribe to five-, seven- or ten-year 'cycles' for wet years. 'We are due for a rain,' some state with confidence. As a result high stock numbers may be maintained when the condition of their country indicates that numbers should be reduced. If the rain does not eventuate or falls on a neighbour's property the stock will loose condition and value. The already stressed vegetation will be overgrazed. Expecting rain in the outback can have dire long-term consequences for the pastoralist, their stock and the country.

Through over a century of heartaches and intermittent successes, most pastoralists in northern South Australia have learned to cope with this unpredictability. They mimic the strategy of many of the animals that have evolved in this erratic climate. Because cattle can't 'do a frog' and confine their activity to periods of peak food availability, they must move to follow the 'green pick'. Just as pelicans follow unknown cues to lakes full of fish, and kangaroos, budgies and emus move to distant areas where rain has fallen, domestic stock must have access to enough land to be able to search out the parts that have received favourable rains.

Some pastoralists acquire huge stations or adjoining properties to increase the likelihood that part of their land receives good rain and can sustain their core breeding stock. Australia's most famous pastoralist, Sir Sydney Kidman, exploited this strategy as he 'drought-proofed' his enterprise with strings of adjacent properties through which cattle could be moved to good feed and markets.

Australian properties are also the largest in the world because of the infertility of our soils. Ted Turner is the largest private landholder in the United States. He runs up to 30,000 bison on parts of his conservatively stocked western ranches, at a stocking rate of over four bison per square kilometre. By contrast, the limit on WMC's 11,360 square kilometres of pastoral leases surrounding Olympic Dam is just over one beast per square kilometre. Actual stock numbers rarely exceed half of this figure to ensure that sensitive or degraded areas are not further damaged. Therefore, even though WMC control over one and a half times the land owned by Ted Turner we can only support a fifth of his stock numbers. But even with such huge properties and low stock numbers, few stations can support their optimal carrying capacity in all seasons. Sometimes the stock have to be moved to greener pastures.

Moving stock around sounds straightforward but it is directly contradictory to the type of land 'ownership' that we have inherited from our European forebears. Under our leasehold system a pastoralist 'rents' a particular block of land for 42 years. In the same way that gold is not taken from another company's mine, farmers do not traditionally graze their animals on someone else's property. Even recent history books are tormented by disasters that attest to the inappropriateness of this 'my patch' mentality with regard to pastoralism in the Australian arid zone.

In the past few decades huge roadtrains, each capable of hauling 160 cattle, have become available to move stock. Faster and more efficient transportation has enabled pastoralists to respond more quickly to deteriorating conditions. Cattle can now be sent to feedlots where food and water are unlimited, or to agistment in areas thousands of kilometres away where conditions are better. Rather than watching their prize breeders perish on their parched land like the pioneers did, we can now sell or move our stock around like nomadic budgies following the seeding button grass. Pastoralists who do not use the 'budgies wings' available to them in the form of the forty-two wheeled roadtrains deserve to face their inevitable extinction.

Despite all of the apparent benefits of a nomadic lifestyle in an unpredictable environment, being constantly on the move also has its drawbacks. Few, if any, animals migrate for the sake of it. If they can make ends meet by staying put they will. In the same way, mustering, trucking and agisting or feedlotting cattle is an expensive, time-consuming business. With ever-increasing diesel costs, there are more and more incentives for pastoralists to minimise their trucking costs. Furthermore, cattle on a new property may have difficulties finding water or may succumb to toxic plants that resident cattle have learned to avoid. It makes sense to modify management practises to limit carting cattle whenever possible.

Adding water points helps to spread the grazing pressure of livestock and enables more stock to be retained at the start of a dry period. Modern earthmoving equipment now enables dams that can hold water for several years to be constructed in a few days. Poly-pipe can relatively cheaply distribute water from bores or large pipelines into previously unwatered country. Land condition has improved largely because we have spread smaller numbers of stock onto more watering points. Dams, bores or waterholes that used to support 1,500 cattle now typically water 150 or less. While these new waters are integral to better management and to reduce the impacts at established or natural waters, they also present perhaps the greatest conundrum for modern land managers.

A few years ago scientists were encouraging pastoralists to spread their waters and stock to avoid Bubbler-like devastations, but now the pendulum has swung the other way. Permanent waters allow stock to be maintained in areas during droughts when they should have been moved away from stressed vegetation. CSIRO scientists are concerned that due to the myriad of permanent waters established in pastoral country, too much of the landscape is now accessible to stock. It is not only the 'feed' that has to be monitored around water points. Some native plants and animals are also sensitive to grazing or trampling by stock, or the spread of weeds or feral animals that may accompany the 'opening-up' of ungrazed country.

The impact of water points is not restricted to introduced stock. Kangaroo numbers have boomed in pastoral areas where water is now freely available and dingo numbers have been reduced. Like goats, rabbits, donkeys and other feral species, increased grazing by kangaroos compounds the environmental effects of sheep and cattle around waters. Additionally, pull up to a dam on a summer's evening and you will probably see hundreds if not thousands of finches, pigeons and parrots. These birds may be eating kilograms of seed each day, seed that would be left to germinate or feed other animals if it were not for the artificially high bird numbers. Crows, eagles and other predators are also attracted to water and in turn may affect the populations of their prey. The relatively straightforward action of installing a new water point results in a whole cascade of environmental effects that change the newly exposed areas.

A balancing act ensues whereby the ability of a pastoralist to maintain an economically viable herd is played off against the sensitivities of different vegetation types and animal communities.

Modern pastoralists must be conservation biologists and play a major role in protecting sensitive and threatened plants and animals on their lease. Through treating their country with the care and respect that it deserves successful pastoralists also safeguard their commercial interests in dry times. It is no surprise that these are the same land managers who are the most outspoken in rejecting ill-founded drought subsidies proposed by the urban media. Financial subsidies or whacky national irrigation schemes will not protect environmental conditions or improve the sustainability of pastoralism. Even small-scale irrigation can be dangerous in central Australia, where rivers drain inwards.

Paul Broad is an entrepreneurial pastoralist who attempted to increase the productivity of Etadunna Station halfway up the Birdsville Track. Part of Paul's strategy was to channel some of his copious energies into ripping hundreds of rabbit warrens. However it was his levelling of land, irrigation and introduction of grasses to fatten cattle that concerned environmentalists. Combined with his diversion of water and the introduction of weeds, the likelihood of increased erosion and salinisation were considered by many people to outweigh the short-term gains that could be made. Additionally, such proposals to irrigate in the catchments of inland rivers could lead to the accumulation of salts and other chemicals, turning our lakes into toxic soups like California's Salton Sea.

Because we are aware of the environmental ramifications of our actions, most outback residents and government agencies oppose agricultural developments in Australia's rangelands. There are even some people and organisations who ask for the entire pastoral industry to be closed down, because they believe it to be out of step with current environmental ideals. To force the pastoralists out would be a huge loss, not only to those of us who enjoy outback hospitality, race meetings and B & S balls, but to conservationists as well. If the pastoralists were forced from the land, who would take responsibility for repairing tracks, controlling feral animals, recording rainfall, manning remote communication bases, supporting permanent settlements, supplies, facilities and infrastructure, and keeping an eye out for unexpected environmental changes? In all of these ways pastoralists enhance conservation work in the outback.

Many pastoralists have a close affinity with 'their' land and a desire to protect the integrity of natural systems. On countless occasions when I have been looking for groves of young pine trees, 'taipan country', following the spread of the calicivirus or the debilitating western myall white-fly outbreak, it is the local pastoralists who have had the answers. These same people have alerted me to occurrences of rare or unusual plants and animals and historical changes in ecological patterns. In many cases their knowledge is similar yet not nearly as well developed as that of the Aboriginal people who have been displaced from much of the rangelands.

I was driving around with a local station manager one day discussing different plants and animals when he made an observation that the spiny *rhagodia* bush was related to canegrass. If he were sitting a botany exam my pastoralist mate would fail dismally. *Rhagodia* is a dicotyledon, a shrub related to bluebushes. Canegrass is a monocotyledon, a grass related to the stuff that they play tennis on at Wimbledon. To a botanist the two plants are at opposite sides of their classification spectrum. But instead of scoffing at the ignorance of the pastoralist, let's look at the two plants through his eyes. Both are long-lived plants that typically reach just over a metre in height. Both grow in temporary swamps, often intermingled among each other and dependent upon intermittent flooding. Neither are particularly palatable to cattle, although they will both be nibbled when green or as a last resort during a prolonged dry period. In terms of landscape and stock production the two plants are similar. They are 'related' yet a laboratory-based botanist would not know this.

On top of this unique understanding of the land pastoralists can help to control one of the main causes of environmental degradation in the outback. Mining, tourism and even pastoralism pale against the overarching effects of feral animals. Although our record is not as dire as some of the Pacific islands, Australia is the continent most seriously affected by feral animals. And of all of our unique landscapes it has been the outback that has been hardest hit by the feral scourge.

Feral predators have caused the extinction of many of our native mammals, and rabbits, goats, donkeys, horses, camels and pigs have surged into the arid zone, wiping out plants, destroying soil and threatening the persistence of many outback species. It is easy to find habitats and whole regions where these feral free-loaders consume more vegetation than domestic stock. Therefore to maximise production while minimising impacts, feral animals must be eliminated or controlled.

When the Williams family purchased Hamilton Station north of Oodnadatta in the mid-1990s they made the removal of large feral herbivores one of their priorities. The man that they selected for the job was none other than Phantom. Phantom is a fast-talking swaggering little bloke with a shock of blonde hair and piercing alert eyes. Whenever I see an advert for a wild gun-slinging western I half expect to see Phantom's names in the credits. Although he is yet to prove his acting skills there is no doubt that Phantom is one of the most colourful and successful feral avengers in the north. On some of my early trips to Dalhousie I had been appalled at the numbers of donkeys, brumbies and camels that were breaking down trees and pugging up waters both on the station land and in the adjacent Witjira National Park.

Within five years of being appointed the manager of Hamilton Station, Phantom had single-handedly accounted for 1,500 of the herbivores that were in direct competition with his boss' cattle and the environment. The change to the creeks and woodlands of Hamilton Station since the mid-1990s is ample evidence that vigilance and persistence in removing feral grazers does work, especially when neighbouring properties work together.

Through my scope I could pick out the tall pale sail-like form of a dark-muzzled bull camel, amazingly camouflaged into the sandy desert for such a huge animal. He flared his nostrils and sniffed at the air, obviously aware of my presence. To the left of him were six camels slightly smaller than their patriarch. I know what Phantom would have done. He would have first dropped the leading matriarch with a head shot, knowing that the rest of the tight-knit herd would stay around so he could finish them all up.

But I did not shoot. In fact my scope was not even mounted on a rifle. I was watching the camels through binoculars and observing what they were eating. While the bull maintained his vigilance, the cows were wandering aimlessly through my field of view. The matriarch grabbed at a leafy sandhill wattle twig with her huge lips and manoeuvred the bright green bundle into her mouth as she sauntered off. Another smaller cow craned her neck down to the ground level between some bushes on the dune and came up chewing what looked like a mouthful of wire. I memorised the spot where she had fed so I could check out exactly what she was feeding on after the herd had moved off. The big bull shook his massive head flashing a bright green eartag. I was not looking at any ordinary feral camels. This was a special herd, the first of their kind.

Phil Gee rode up on his motorbike with his expansive cameleer's beard parted down the middle and comically flowing across both shoulders. 'They have split into two small mobs,' he said and then continued with an air of relief, 'The other mob has walked along the fence up near Coomandook, but they're all still in this paddock.' I was relieved too. My two main reservations with allowing Phil to embark on his passion for commercial camel grazing on one of WMC's stations were that the camels would either seriously damage the vegetation or they would destroy our fences.

In December 1998 Phil received approval from the Pastoral Management Branch to trial camel pastoralism on Stuart's Creek Station. A year earlier I had taken over the responsibility of managing Stuart's Creek and three other stations owned by WMC in the district. Phil was lured by the lucrative markets for camel meat in the Muslim countries of the Middle East and South-East Asia. In addition he saw great potential in developing markets for dried camel milk and for tasty steaks in countries gripped by 'mad-cow' hysteria. According to Phil and the Central Australian Camel Industry Association, of which he was a member, the main hurdle for the budding Australian camel industry was to ensure a regular supply for their markets. Mustering camels from remote feral herds could not guarantee that the industry could meet orders promptly and efficiently. Unlike feral goats that could be more efficiently mustered by sheep owners down south, the camel marketers needed to have domesticated camels on hand so that they could respond efficiently to orders.

It was the pastoral pioneer Sir Thomas Elder who first introduced camels into South Australia nearly 150 years ago. Elder did not take camels to his Beltana Station for pastoral purposes but rather to 'open up' the vast interiors through the camels' phenomenal transportation capabilities. Up to 12,000 camels are thought to have been imported to Australia and many more were bred at stations like Muloorina before their haulage duties were replaced by trains and trucks. Now camel rides at the occasional tourist park, camel treks for the more adventurous travellers and highly entertaining camel races are the only reminders of the importance of camels in pioneering days.

I was interested in supporting a camel grazing trial because of the potential environmental benefits of co-grazing camels with cattle. Camels browse predominantly on vegetation that is of low value to cattle, who predominantly eat grass. When I checked out the 'wire' that I had seen the camel eat I found that it was the tangled, prickly caltrop creeper. This annual weed, which flourishes in sand dunes disturbed by stock or rabbits, possesses the most horrendous prickles imaginable. Andrew Phillips from the Department of Primary Industry and Fisheries has also found that camels in the Northern Territory feed on other prickly nightmares, like galvanised burr, goathead and buckbush, as well as browsing on acacias. Because they eat as they walk, low numbers of camels typically cause little damage to trees because they only take a mouthful or two before moving on. Likewise, their large padded feet do not break up or compact the soil surface to the same degree as hard-hoofed stock. Together with the South Australian Pastoral Management Branch I installed a series of vegetation and soil-monitoring sites to determine what effects Phil's small herd would have. Two years later we could hardly even see where the camels had been.

Their long legs and iron digestive system, along with their famous ability to survive for long periods without water, make the camel one of the world's premier desert-adapted mammals. It is little wonder that disease-free feral camel herds have thrived in unwatered Australian deserts where most cattle succumb to their first drought. These same attributes that benefit feral camels can also benefit some progressive pastoralists. Because they wander further and water less regularly camels require far less watering points than cattle. Fewer water points save pastoralists both money and time and also reduce the effect on the environment.

In some preliminary studies north-east of Alice Springs, Andrew Phillips found that steers that grazed with camels gained weight slightly faster than the same density of steers isolated from camels. It is possible that the steers accumulated some of the camel's gut flora, which allowed them to digest more of the leaves and twigs that they ate. Presumably, if the numbers of steers were reduced when camels were introduced to their herd they would perform even better.

Perhaps because they have experience with feral camels, many cattle breeders are concerned that camels will either stir up their cattle or destroy their fences. Andrew reported no aggressive interactions with cattle even when camels and cattle were held in the same yard.

Phil has also found that after only a few days of 'training' feral camels to respect yards and fences they are no more likely to knock down a fence than domesticated cattle.

Camels offer some exciting possibilities for relatively environmentally friendly grazing in some parts of the arid zone. However they too have their drawbacks, from both economic and environmental perspectives. Camel cows generally only produce one calf every two years and hence the herd takes twice as long to breed up compared to cattle. Furthermore, because they are so tall, camels cannot be carted in double-decker trucks. As a result, transport costs that are becoming more restrictive as fuel prices increase are considerably higher for camels than they are for cattle. Bulls may also fight with each other and therefore need to be separated. However, the mounting evidence suggests that where small mobs of camels are run with reduced cattle herds, pastoralists can lessen their environmental impacts, increase beef production per head and tap into a potentially lucrative market.

Of course, like any introduced animal, camels invariably have a detrimental affect on the environment. In places where large numbers of feral camels congregate, favoured trees such as the elegant wattle and quandong are particularly hard hit. Goats are now used as valuable diversification options for some sheep graziers, but they also trash the vegetation and soils of hilly country, probably to a greater degree than camels. Although in some environments camels and goats are more favourable from an environmental or economical perspective than cattle and sheep, they are not the optimum grazing animals for the Australian rangelands. Indeed, the best methods of minimising the environmental legacies of pastoral production throughout the outback have yet to be determined.

23

Gengelopes

I am sitting on a rolled-up swag staring into the orange tongues of flame that lap from the old Ghan railway sleeper into the dark night air. Empty beer and Coke cans glint from the ring of swags on my side of the fire. Over to my left is a group sitting on trendy deck chairs, others on my right are perched on their eskies. A small party of students who forgot to bring chairs pulled the bench seat out of their Kingswood. I can just make out several bare-footed hippies sitting cross-legged on a rattan mat, drinking herbal tea. Their dog wanders over and sniffs a baby lying on an Australian Geographic rug, inviting a rum drinker to crack an Azaria joke. The baby's dad, who is wearing white socks, quickly puts the kid in the back seat of his Pajero.

The conversation bounces backwards and forwards across the fire. Everyone contributes when their area of expertise is brought up then sits back listening intently to the flow of ideas and concepts. The dialogue is informal yet captivating. In keeping with any meeting where everyone has something to contribute and nothing to lose, no-one is big noting or point scoring. I hear no spiteful innuendos or suspicious interrogations.

Although some of the faces and voices are familiar, most are not, which is surprising since I had arranged for them all to meet tonight. I suspect that the bloke in the white socks is the marketing guru with whom I had been put in contact. One of the women that he brought from Sydney is apparently Australia's leading food nutritionist and was recommended to me by both the National Medical Council and the heart-disease crowd. The marketer's wife was also invited as she was a public opinion pollster with the uncanny knack of predicting what we would clamber for in the future. She in turn had put me in touch with the crusaders on the rattan mat who were delighted to be involved but who were not so keen on the bush-butchered steaks.

Among the contingent of carefully selected scientists and land managers on the swags was Peter Latz. Latz was eating wild food when the 'Bush-tucker Man' was in nappies, and he had driven down from Alice to contribute. Brian Powell from Quorn, who has grown almost every type of plant in the outback with the aid of a bucket of salty water, had come along also. He was personally responsible for planting many of the trees that now shade towns and homesteads in outback South Australia, and was rightfully proud of his borewater-irrigated orchard and vineyard. Through years of careful cultivation Brian had developed large juicy quandongs that were now the core of a lucrative and expanding market. If anyone knew about edible native plants it was Latz; if anyone knew how to develop wild plants into commercial crops it was Brian. Next to Brian was Elliot Price. I had never met Elliot but I had viewed his grave overlooking the waterhole at the Muloorina Homestead he built near Lake Eyre. Elliot was more than a pioneering pastoralist, he was also an inventor and an innovator, reputedly decades ahead of his time.

Another blast from the past was Brigitte Bardot. Although I could have rounded up a better-credentialed animal rights campaigner, Brigitte's name had cropped up and, like most of the blokes around the fire, I was pleasantly surprised that she accepted my invitation.

We had all gathered around the fire to brainstorm the fate of Australia's rangelands given the recent advances in genetic engineering. Genetic engineering had been touted by its protagonists as the answer to many of our environmental problems. Crops or farm animals could have their genetic blueprints altered to render them more productive. These engineered varieties supposedly require less pesticides, fertilisers or dietary supplements than ever before. Amazingly, genetic engineers have frost-proofed tobacco and tomato plants by inserting into them the antifreeze proteins from the blood of polar fishes. There is already talk about manufacturing new viruses to control cane toads or foxes. The potential scenarios for this technology are seemingly limitless. I was particularly interested in the effect of genetic engineering on the type of production we would witness in the Australian outback in decades to come. Will we have onion-flavoured spinifex and apples growing on wattle trees? Maybe we will be harvesting cucumber-sized witchetty-grubs for the Asian export market, or an assortment of flavours of jug-sized honeypot ants for French patisseries!

All of us had a stake and a genuine interest in developing the ultimate outback plant or animal that would provide the best food source for our ever-increasing population. Our collective task was to conceptualise the single organism that would best suit the interests of the bushies, the environmentalists and the big-city consumers. Our prototype would enable the genetic engineers to develop a product that was compatible with the environment, could be harvested with minimal use of energy and infrastructure and that was desirable to consumers, whether their motivations were health, taste, prevention of cruelty or resource sustainability.

The first thing we needed to decide upon was whether the future of rangeland production lay in crops or meat production. Crops of any kind require relatively reliable rainfall. I had made an executive decision not to invite a meteorologist to the brainstorming. Until scientists can extract genetic material from waterfalls for inserting into blue sky, central Australian producers must contend with an unpredictable and generally dry climate. The little CSIRO soil scientist who was wearing brown pants and a brown jumper piped up and added that inadequate rainfall was not the only hindrance to broadscale agriculture in the outback. Through a mischievous grin he said that until the genetic engineers can augment infertile sand grains with DNA from chook manure, there was little value in considering cropping because our soils are not fertile enough. Surprisingly Brian Powell agreed. I had invited him to press the case for super-crops or fruits derived from native species, but he agreed that these could only be a feasible sideline in small irrigated areas.

The swag sitters relaxed and the mat squatters squirmed as I made a decree that due to our climate and soils we would have to focus on animal, rather than vegetable production. We were going to start from scratch to develop a genetically engineered animal ... a g.eng. critter ... a gengelope.

In the interests of keeping everyone happy those who had preferred a vegetable option were asked to initiate the development by setting their preferences and conditions for the gengelope. They almost clambered over each other to compile their wish list. The couple in the cloud of smoke behind the Combie raced over to their mates on the mat and the students sprang out of their bench seat where they were indulging in a Coopers Ale drinking competition. Cruelty to animals was a huge issue for both groups. They acknowledged that livestock were now treated much better due to contemporary stock care initiatives, but required that our gengelope had minimal handling or treatment. Dehorning, mulesing and castration were ruled out; they even outlawed branding and ear-tagging as options. A gengelope would have to be taken to markets without being pushed into a roadtrain and must be slaughtered in a way that it felt no pain or stress. Brigitte positively beamed as the developing list appeased most of her concerns.

Now the swag sitters were getting a bit edgy and were strengthening their rums. They could not conceive of any feasible way that they could be involved in primary production in the outback within these guidelines. However, the public relations expert and her marketing partner were nodding as the conditions were being laid down. If gengelopes were to appeal to all sectors of the public, all of these traits would be desirable. I was concerned that I was losing the resolve of my mates on the swags but reminded them that genetic engineering is an amazing technology and that their turn for input would come soon.

To continue the wish list the nutritionist was advised to put in her order. Fortunately she refrained from delivering a speech on obesity, heart disease and a balanced diet, but she did indicate that gengelope meat should be lean, low in cholesterol and rich in protein. We were instructed that the gengelope should not be exposed to any hormones, antibiotics or substances that could affect consumers' health or perceptions. A hygiene inspector who had tagged along with her in the Pajero chipped in with a clause about eliminating the chance of a 'mad gengelope' or foot-and-mouth disease. Mr Whitesocks, the well-restrained marketing man, was chomping at the bit for his time in the spotlight. 'Unless the meat was tender, tasty and easy to cook,' he stated with authority, 'we are wasting our time.'

A quaintly spoken little fella with a manicured hairstyle introduced himself as a 'textile engineer'. Although none of us had ever heard of such a profession we listened to his authoritative comments. The agricultural regions apparently produced enough wool for the world's markets. He wanted us to incorporate a new designer leather into our gengelope prototype. Like a leather version of titanium the textile engineer wanted the hide of the gengelope to be lightweight, strong, supple and tear proof.

The stakes had increased considerably and I figured that it was an appropriate time to ask the pastoralists for their wish list, before the rum took over. Although they acknowledged that the make up of the gengelope was primarily driven by market forces, they were caught up in the spirit of opportunity and came up with some demands of their own. The gengelope should be able to thrive on native pastures and untreated waters. Gengelopes should be large, wide-ranging and mobile enough to take advantage of patchy rainfall. Elliot quipped that a gengelope would therefore have to be a far cry from a sheep, which cracked everyone up except the hygiene inspector from New Zealand who seemed to miss the humour. Distressed by the loss of lambs, calves and breeding stock in dry times, the pastoralists wanted the gengelope to coincide its breeding efforts with sporadically favourable conditions. A gengelope should require minimal management, monitoring, fencing and specialist infrastructure and should be resistant to disease and flystrike. If the genetic engineers could fit these requests while keeping the sensitive consumers happy, the pastoralists vowed to embrace the gengelope with open arms.

But we were not finished yet. The World Wide Fund for Nature representative, who had strategically placed himself between the pastoralists and the vegetarians, could only endorse the gengelope if its production would not threaten any native plants or animals. This meant that the gengelopes could not graze preferentially on seedling trees like rabbits and some stock do.

He was also keen for the gengelope to have an efficient gait and padded feet like a camel so that it did not destroy the soil crust.

Just when I thought that all of the stakeholders had said their bit, the little soil boffin spoke up again, throwing in another curveball. He was still concerned about our soils and pointed out that any sort of harvesting for the rangelands would contribute to the depletion of scarce soil nutrients. 'Every time we move an animal grown from the rangelands to the markets in the coastal fringe of Australia,' he explained, 'we are literally "mining" the minerals and nutrients from the desert soils.' What was worse was that the nutrients locked up in the offcuts and offal, heads and hooves contribute to further environmental problems in landfill or ocean outflow in the major cities. He wanted the wastes of the gengelope to be returned to the inland where they were so desperately needed.

I asked my graphic artist and lateral-thinking mate Dave Kovac to come up with a master plan of the gengelope to assist the DNA jockeys with their mission. These genetic engineers had been scribbling down the wish list and looking more and more fazed, suspecting that we had pushed their technology beyond its limits. However, in my dream world there are no limits. So I threw the cap from a celebratory bottle of rum and toasted my happily buzzing companions. Our group would never be able to reach consensus about a favoured flag design, Head of State or preferred beer yet within a few hours we had designed an entirely new animal and potentially reshaped the future of Australian diets and pastoralism.

The next time I was sitting around a fire I could feel its heat and smell the sweet smoke of native pine. This time my mind was not drifting, yet I was watching the fire intently. It was not the hypnotic flames that I was staring at, but the singed carcass that had been thrown onto the coals. All of a sudden it struck me that maybe, for the first time, I was looking at a pretty reasonable prototype of a gengelope.

A few times a year WMC collected a group of senior Aboriginal men from north of Coober Pedy to seek heritage clearance for planned developments. These old blokes, some of whom had worked on the local stations as lads, tracked dreaming trails and advised on the cultural significance of various claypans, rock outcrops or stone arrangements. Construction of roads, pipelines, drill pads, dams and powerlines was subsequently planned to avoid sites of particular significance.

Jimmy Bannington was one of the regulars on these visits and of all of the men he was the most concerned about their diet. Jimmy's ample belly attested to years of indulgence. While the other men were discussing the importance of particular sites, Jimmy would invariably swagger over to me and rumble between closed lips 'Camp tucker wiya, malu tonight, palya?' So that evening I would take Jimmy and his mates out to shoot a kangaroo. But not just any kangaroo. Jimmy and the others would carry on like a carload of stock inspectors discussing the relative merits of a herd of cattle. 'Dis fella one ere', 'Wiya, dis one', 'Here, here, here, good malu, fat one here, dis one!', 'Palya?', 'Uwa'. Eventually they made their decision and selected their tucker.

While others in the 'heritage' party were dining on the limitless camp food prepared barely a kilometre away the old fellas made their own dinner. The gut of the roo was pulled out of a small slit in the belly. They then stitched up the slit with a stick that was held in place by wrapping the intestine around both ends. A minute or so on both sides in the flaming fire quickly singed off most of the fur, drowning the sweet smell of pine smoke. According to custom, the tail and lower legs were removed and the tail was laid next to the carcass in the fire that had burnt down to a bed of coals.

The men were always talkative and content when cooking themselves a roo. Apparently arbitrarily, when one of their stories had finished, the roo was flipped over. I timed the whole cooking process at thirty-five minutes, but I was definitely more interested in the time factor than the chefs were.

While the cooking appeared to be casual, the carving was a classic.

With a few deft slices of a shovel, chunks of steaming flesh and charred hide were separated, inadvertently receiving a light garnishing of sand and ash. Willy Dingle appeared to be particularly fond of the tail. I followed his lead and hacked off a segment of tail and removed the singed skin, like peeling a hot banana. After we had finished with the tasty tail meat, Jimmy handed me the sweetest and juiciest red meat that I had ever tasted. Each of the men had their favourite portions, but I had no idea where my irregular-shaped hunk of meat had originated. It was rarer than rare and so tender that I barely had to chew.

While the others were noisily devouring their favourite cuts, I thought back to what Mr Whitesocks had said about the gengelope being tender, tasty and easy to cook. This roo definitely met his criteria, but my current experience was probably not the mass media marketing approach that he would have envisaged. Irrespective of the finer details I wondered how well the roo on the fire conformed to the other attributes of the gengelope that we had brainstormed.

Should we add the roo's tail to the gengelope prototype? Apparently, when they are travelling at speeds of greater than twenty-two kilometres per hour, the energy expended by a hopping animal with a counter-balancing tail is less than that of a running animal like a sheep or cow. Red kangaroos are known to travel up to 300 kilometres to concentrate in areas that have received rain. Clearly a long tail would assist the gengelope to cover the long distances required to locate suitable food and water in the outback. But the kangaroo could still offer more suggestions to the gengelope-design team. Because their metabolic rate is thirty percent lower than sheep, kangaroos have lower food and water requirements than conventional stock. Gordon Grigg, kangaroo researcher from the University of Queensland, has calculated that due to their low-energy requirements and smaller size a kangaroo eats less than one fifth as much as a sheep does. I made a mental note that a low metabolic rate was another feature to be incorporated into the gengelope design.

I picked up one of the discarded feet. It had soft pads and had been connected to the leg by incredibly strong shock-absorbing tendons. Both the pad and the tendon reduced the kangaroo's impact on the ground compared to hard hooves. The tendons also spring the leg back to its starting position while hopping, minimising the energy sapping use of muscles. Furthermore, by hopping kangaroos don't drag their feet as much as sheep, thus reducing their impact on the soil and vegetation. It was becoming apparent that one of the staple foods for Aboriginal people for thousands of years was already upstaging our gengelope in several design features.

I know that roo meat is cholesterol free and is squeaky clean of growth hormones, herbicides or even genetically engineered tainting. But is it hygienically clean? This took on a real relevance when I noticed my bloodstained hands in the flickering firelight. Several old-timers don't eat roo meat because it is 'full of worms'. The following day, without mentioning the word gengelope, I rang the Australian Quarantine Industry Services (AQIS). Tony Weeks works for AQIS, which is responsible for clearing Australian meat products for export to fussy international consumers. His counterparts in the state meat hygiene departments perform a similar role for any meat produced for local consumption. When I asked him about the hygiene attributes of kangaroo meat Tony was initially defensive. 'It is a really healthy meat, low fat and ... ' I cut him short. I knew it was healthy but I wanted to know whether consumers should be concerned about worms or other nasties that are sometimes associated with game meats.

Tony explained that since 1997 there has been an Australian Standard for the production of kangaroo meat for human consumption. It is the responsibility of the meat producers to adhere to the guidelines, which were policed by the meat hygiene officers. Since this standard had been introduced all meat sold for human consumption was 'safe', 'clean' or whatever you wanted to call it. In fact the kangaroo industry claims that the rejection rate for kangaroo carcasses is considerably less than that for sheep or beef. Tony stressed that whether the meat is pork, beef, chicken, roo or lamb, if it was sold at a butcher or supermarket, consumers had the same guarantee that it was free of nasties.

That was comforting, I admitted, but I would not let him off the hook about worms. 'Kangaroos have worms don't they?' I challenged, 'Doesn't that increase the risk to roo consumers?' Tony replied that each meat industry had its own 'challenges' that they had to manage in order to satisfy the hygiene regulations. Yes, some kangaroos from certain areas carried stifle worms that rendered the meat unsuitable for human consumption. These worms, which normally live out of harm's way low down in the leg, were easily detected and the field processors were instructed not to harvest such animals. The field processors benefit from identifying and isolating these animals because they cannot sell infested meat. As a result the roo meat presented to the hygiene inspectors is free from worms and the product that turns up in the fridge is absolutely clean.

With the quality of the meat beyond reproach I next considered the attributes of roo hide. How does kangaroo skin shape up to the strong, light and flexible hide that the textile engineer had ordered? Rita Borda is the outspoken chairperson of the Kangaroo Industry Reference Group and also the manager of Macro Meats, South Australia's largest kangaroo-meat exporter. With her bushfire of claret-red hair trussed up into a riotous arrangement and her passion for the kangaroo industry, I regularly see Rita extolling the virtue of kangaroo products to sceptical or ambivalent audiences. Rita's response to my question was instinctive, like she had slotted in her 'hide' tape rather than her 'health', 'cleanliness' or 'humane' tape. I was told that kangaroo hide is renowned for its high tensile strength and low weight. It is the titanium of hides. Apparently the world's most famous soccer player, Maradonna, insisted that his soccer boots were made of roo leather. In addition to being a light, strong and supple leather for sports shoes, kangaroo hide is also widely used for ladies footwear and souvenirs. Like the meat, the hide of kangaroos was beyond reproach. What about the other factors raised by the gengelope committee?

One of the key concerns of the brainstormers was the cruelty of harvesting gengelopes, so I spent a few nights with a roo shooter, or 'field processor' as he is now called, to learn first hand. I met Lindsay King after dark. According to their regulations, field processors must commence after dark and have all of their carcasses loaded into chillers before dawn. Lindsay is a no frills ex-navy and railways man, who is more than a bit rough around his ample edges. By complete contrast his Landcruiser utility was startlingly clean, unnaturally so given that it rarely drives on bitumen and regularly carries more than a tonne of kangaroo carcasses. I almost felt like I should clean my boots before climbing into the passenger seat where I carefully nursed his prized .223 rifle. The gun was even better maintained than his vehicle, for good reason.

Soon after we headed off, the powerful roof-mounted spotlight picked out three roos grazing in a small grassy patch. 'Keep it just in front of the little harlot on the left. If she stops those two big pricks will stop too', Lindsay barked as he propped his rifle on the padded bar outside his window.

Bang. The head of the doe exploded and recoiled backwards.

Bang, reload, bang.

The two larger buck roos that I had been 'holding' just out of the intense beam dropped where they had been grazing. If his rifle's scope had been out by only a fraction Lindsay would have either missed the roos, or worse still, shot them in the neck. Although the chest presents a larger target than the head, the RSPCA-endorsed protocol for kangaroo harvesting stipulates that they must be shot in the head. Head shooting is reinforced by kangaroo processors who downgrade or do not accept carcasses that are not shot correctly. These regulations mean that unless the roo is shot in the head it is a waste of a life, a bullet and a carcass worth about ten dollars to a field processor.

We drove over to the three fallen roos. The female had a hairless joey in her pouch that was quickly dispatched before the roo was heaved up to a hook on the outside of the shiny steel tray. 'She must be on the move, see', Lindsay explained pointing to the dirty pouch. 'She hasn't had time to clean it out.' Doe kangaroos that have been living in the same region for a while normally keep their pouches as clean as Lindsay's tray.

I struggled to hook one of the bucks that probably weighed just over forty kilos onto the side of the tray. Although I could lift it high enough, I could not simultaneously bearhug the huge beast and direct its leg to the hook. Lindsay laughed as he gave me a hand with a well-practised lift and lunge. 'That's nothing, wait until we get a sixty-kilo prick, that'll sort you out!'

Hanging them by one back leg, Lindsay slit the throats of the carcasses to bleed them straight away. When we had a few more roos hanging on the outside we stopped to 'knock their fluff off'. With experienced gloved hands and knives that had been sharpened to thin curved blades, Lindsay removed their guts. The heart, liver, kidneys and lungs were left in the carcass for subsequent inspection by the meat processors and AQIS officials. My job was to lop off the head and tail, clip off the feet with the largest pair of bolt cutters that I had ever seen, then insert the numbered and dated tag into a slit at the base of the tail. This tag allows National Parks officers, meat processors and hygiene inspectors to identify where and when the kangaroo was 'harvested'.

Within a couple of minutes we were looking for more roos. The roos that were picked out by the sweeping spotlight were typically unfussed by our approach. On some occasions Lindsay had to beep his horn or whistle to distract them from their feeding so they would lift their heads to enable a clean shot. The kangaroos that Lindsay shot while I was with him typically suffered no stress or pain. But of course I have only seen Lindsay shoot his nightly tally of forty or fifty roos on a handful of occasions. I wonder how the roos harvested by other field processors fared?

One bloke who should know is Dr Martin Denny, a kangaroo biologist who was commissioned by the RSPCA to assess the compliance of commercial shooters with their code of practice. In 2000 Martin painstakingly investigated thousands of carcasses harvested by professional field processors. Whereas a similar study fifteen-years earlier showed that fifteen percent of the harvested roos were not head shot, Martin found that this figure had dramatically improved to only three percent and that the vast majority of these non-conformers were shot in the top of the neck. It is not surprising that Martin concluded that the professional shooters were incredibly accurate marksmen. He was confident that the proportion of neck shots would reduce even further now that field processors understood that these were no longer acceptable. He even went further to say that professional shooting was the most humane form of death for a roo, outstripping being shot as pests by less accurate non-professionals or dying of thirst or starvation. I bet that domestic stock waiting in an abattoir's yard wish that their life had ended at the hands of a field processor.

Kangaroos are one of the exceedingly rare examples of commercially available game species that spend their entire lives as a truly wild animal. Although harvesting temporarily disrupts the kangaroo 'society', and the consistent occurrence of joeys in the pouches of females is an unfortunate result of their efficient reproductive system, I struggle to imagine a less stressful or intrusive method of harvesting meat. More than any other animal on this planet, wild harvested kangaroos satisfied the gengelope's strict cruelty stipulations.

I was surprised that Lindsay discarded the tails since they are a delicacy in some Aboriginal communities. But there is only a limited market for roo tails and in most cases they were not economical for the field processors to collect, chill and transport. As I drove around with Lindsay I lamented the waste but realised that leaving the tails in the bush, along with the guts, heads and feet, has its benefits. Just as the soil scientist had requested in my gengelope dream, these offcuts that amount to nearly a third of the kangaroo's weight, do the bush a favour when they are converted back to essential nutrients by scavengers and decomposers. On one occasion I counted sixteen wedgies on a dump of roo entrails and offcuts. I'm sure that they too appreciated the shooters leaving them a share, especially when rabbit numbers were low.

Given that kangaroos seemed to conform to the gengelope that had been brainstormed, I did more research to determine whether they could become a viable and appropriate diversification option for the rangelands. Approximately 3,000,000 kangaroos are harvested annually in Australia, while kangaroo numbers continue to rise in many areas. There are plenty of kangaroos but is their harvesting economical? Not surprisingly, given the low price paid to producers of kangaroo products, David Croft from the University of New South Wales calculated that at $1.50 per kilogram, kangaroo harvesting was only viable for producers in those years when sheep and wool prices were low. An accountant with tunnel vision would therefore advise a pastoralist to reduce kangaroo numbers and maximise sheep numbers. But let's broaden the equation.

Ten years ago only about five percent of the harvested kangaroo meat was used for human consumption and South Australia was the only state that embraced our national meat. Today, roo meat is sold for human consumption in all states and nearly a quarter of all roo meat reaches this high-value market. With an increasing proportion of the harvest going to the human-consumption market, the returns from roo harvesting will continue to improve. What if kangaroo meat attracted a large share of the market for domestic consumption at a price comparable with beef or lamb, as its quality demands? What if the educated public and lucrative export markets were prepared to pay a premium for the most humanely harvested, healthy meat on the planet? Roo meat already retails for more than $20 per kilo in Sydney. The $1.50 per kilo for roo meat that David Croft used in his equation could more than quadruple. And that is just one side of the ledger.

'Farming' kangaroos is impossible and unnecessary. Conventional sheep and cattle fences offer no barrier at all to kangaroos. Fences are required to isolate sheep and cattle from neighbour's stock and to spread them appropriately to manage grazing pressure. However, kangaroos don't 'belong' to anyone and they naturally disperse to the areas with the best food and water. Therefore landholders who harvest kangaroos commercially, and reduce the numbers of conventional stock accordingly, are 'rewarded' by roos immigrating to their land if they provide a better environment than neighbouring properties. Fences that are costly and time consuming to build and maintain become largely redundant in regions where kangaroo harvesting is the focus. And its not only fences that are not required. Imagine the benefits of not having to build or maintain yards and shearing sheds and not having to check water points every day in summer.

A greater emphasis on roo harvesting and a concomitant reduction in conventional stock levels will improve the long-term sustainability of the country. More scarce nutrients will be recycled locally, the grazing pressure will be more natural and soil compaction will be reduced. Damaging high kangaroo numbers will be averted once kangaroos become a more valuable commodity, rather than a pest. Everybody wins.

Everybody may win but do the roos win? What about the conservation and genetics of kangaroos, assuming that ten to twenty percent of the population are shot each year? If the field processors are only permitted to harvest a very limited number of roos each year it is logical to expect that they would select the largest, most valuable males, which in turn may artificially reduce the size of adult males in the population. State Government environmental managers set annual kangaroo-harvesting quotas on a district-by-district basis according to population trends and prevailing conditions. However, these conservative quotas are seldom if ever met. Field processors are generally not limited by the number of roos that they are allowed to harvest. Rather, as Lindsay King points out, it is more profitable and efficient to harvest roos as they are encountered, rather than burning up fuel, time and tyres searching for only the biggest bucks. Indeed, over the previous three years, nearly two-thirds of the 14,000 roos that Lindsay has harvested locally have been females, which is roughly the same percentage as they those that occur in the wild. He is clearly culling the population representatively rather than placing artificial selection processes on it.

All of the arguments for kangaroo harvesting have convinced many of Australia's most eminent environmental scientists that it is an industry that should be promoted. Along with long-term kangaroo researcher Gordon Grigg, eminent museum curators and naturalists Mike Archer and Tim Flannery have been extolling the benefits of regulated kangaroo harvesting for years. They have been calling on Australians to reduce the levels of conventional stock and to increase the domestic consumption of kangaroos. Proper management of kangaroo numbers will, they argue, assist the environment and the kangaroos. It makes so much sense.

A greater emphasis on kangaroo harvesting makes sense but will it work? A similar move was attempted in southern and eastern Africa in the 1960s when a movement was initiated to replace some of the introduced cattle herds with native eland. Eland are a large docile antelope that can be herded and that have a reasonable quality meat. However, cattle are the traditional banks of wealth in these countries and the huge European market was for beef, not eland. Sadly this venture failed. While the kangaroo industry does not have to contend with strong cultural ties to cattle and the Australian domestic market is large enough to create considerable demand without relying on overseas markets, they do have to contend with the *Skippy* syndrome. Many Australians have difficulties rationalising that a cute native animal can also be a pest or a resource.

Slowly but surely Australians are overcoming their aversion to eating our national symbol. Gradually the negative vibes of some ill-informed animal liberationists are being overcome. Very few of us grew up with roo on the table, yet year by year the number of Australians embracing kangaroo meat is growing. For many who previously ate it as an occasional change from 'conventional' red meats it is now their meat of choice. Others who baulked at the idea a few years ago are now trying a garlic roo roast, a marinated kebab, or using kangaroo mince in a lasagne for the first time.

Perhaps the most unlikely developing market for roo meat is not the red-blooded meat'n'three-veg types, but the vegetarians. Yes, that's right, some 'moral vegetarians' are jumping on the bandwagon as well. It is an enlightening experience to witness the anguish as some vegetarians come to the realisation that eating a roo steak is better for the environment than eating lentils or rice, which required cleared land, irrigation and possibly a host of chemicals. For the rest of us who are prepared to eat battery-hen chickens or fish that must experience considerable stress no matter how they are caught and killed, the inclusion into our diets of free-range kangaroos is perfectly rational.

With even a little more lateral thinking, planning and education we could also be harvesting other native gengelopes. Rather than waiting for the white-coated genetic engineers to come up with solutions to rangeland diversification, we should continue to re-evaluate our ethics and conservation management plans. Crested pigeons and emus immediately spring to mind. Both have increased in numbers to unnatural levels since waters have become widespread and permanent throughout the rangelands. At times emu populations build up to the extent that they damage fences and the environment, and they are culled and left to rot in the paddocks like donkeys and camels. The commercial harvesting of emus is currently conducted in farms, which are ecologically and ethically equivalent to big chook runs. It would be more humane for the emus and more beneficial to the environment if free-range emus were harvested on occasions.

Kangaroos and other native animals are not the concoction of an imaginary assortment of outback dreamers assisted by the scary new science of genetic engineering. They are one hundred percent natural and quintessentially Australian. While a Union Jackless flag and an Australian Head of State are regularly touted as milestones that are necessary to prove that Australians have truly come of age, a more tangible measure of our originality and independence will be when our national airline serves kangaroo, instead of battery-hen chickens or other inappropriate reminders of colonialism.

PART SIX
Fighting off the ferals

24

More than meets the eye

The Cobbler Sandhills are one of the most desolate places in Australia. If you had to picture a whitewashed Armageddon, the Cobblers would be a good starting point. By comparison the Dismal and Moon plains of plains rat fame appear luxuriant. At least these plains are still pretty much in their natural state, with most of their plant and animal communities intact. The Cobblers, and indeed the entire Strzelecki Desert between Lyndhurst and Moomba, has been absolutely trashed. The only green leaves too high for marauding rabbits to reach are on scattered acacias or the stunted coolabah trees that mark the flood-outs or tributaries of the Strzelecki Creek. Just like the Bubbler region, the Cobblers was transformed from a productive area into a wind-blown wasteland by early pastoral disasters and rabbits.

Despite its eroded desolate appearance the Cobblers still holds a few wildlife gems. For a herpetologist the prize of the Cobblers is the elusive woma python. Unlike the smaller, boldly banded womas found in the spinifex deserts of Western Australia, the womas from northern South Australia are relatively unmarked and absolutely enormous. The first of these huge snakes that I found was, at the time, the largest snake I had ever attempted to catch.

I was driving up the Strzelecki Track with Fewster, another mate called John and his young son Jamie. We had stopped in an unremarkable part of the Cobblers when not twenty metres from the car a massive yellowish apparition melted away behind a mound of hard white sand. I screamed out 'WOMA' at the top of my lungs and raced after it. I stopped the snake by standing in front of its rabbit warren refuge and had one of those 'Yeah!' experiences that an athlete celebrates when they break a world record. The others had no idea what a woma was but my yelping and whooping soon had them running over to see what all the excitement was about.

To measure and photograph the rare snake I first had to catch it. Grabbing a non-venomous snake is simple, especially when its tail is well over two metres from its head and its reflexes are understandably slowed by its bulk. Hanging on to it and stretching it out to get an accurate measurement is a different story. The huge snake had not shown any sign of aggression when I cornered it and then grabbed its tail. But as soon as I held its head the woma decided that enough was enough. The monster hissed its disapproval and my hand holding its neck was involuntarily dragged behind my back, while my other hand was pinned to my leg by the woma's muscular tail. Within seconds, the snake was beating me at an enforced game of Twister. I was powerless to control the stroppy woma that was threatening to dislocate my shoulder. The snake could have tightened an iron-strong coil around my chest or neck and suffocated me as it does its prey.

Reluctantly, I had to release the woma's head, which gave me even less control and potentially exposed my face to its sharp teeth. Fortunately womas are not man-killers and this one did not bear a grudge. As soon as its head was free the python relaxed. Within a few minutes the two-and-a-half-metre giant that was as wide as a foam stubby-holder was draped placidly over young Jamie's neck with his proud father snapping happily on his camera.

Womas are spectacular, endearing and possibly endangered animals. The monster in the Cobblers was the highlight of that weekend, but womas are not the rare wildlife that usually lures me up the Strzelecki Track. The typical quarry for a group of National Parks biologists and occasionally their friends from Roxby were hopping mice. Although they hop like miniature kangaroos and closely resemble the marsupial kultarr, hopping mice are rodents. Ten species of hopping mice, or *Notomys*, used to grace Australia's deserts. Now only half of those remain and several of the surviving hoppers are on tenterhooks.

Perhaps the rarest—the dusky hopping mouse—was once widespread throughout the central deserts, but for now it is apparently confined to a few dune areas including the Cobbler Sandhills. Way back in 1923 when writing about dusky hopping mice, the mammal curator for the South Australian Museum—Frederick Wood-Jones—lamented that 'without patient study we shall not attain any real knowledge of our smaller mammals before the day of their inevitable extinction arrives'. Katherine was one of the biologists who were charged with the responsibility of discovering where the last of the populations of dusky hopping mice lived and researching their biology. Her job was to prevent the fate that Wood-Jones had predicted.

There has to be something special about the Cobblers that enabled these rare mice to persist there, while disappearing from most other places. Exactly what made these sandy deserts so special was one of the most puzzling mysteries of the outback. Even more amazing is that one of the biggest and certainly the best-studied populations of these hopping mice is found right next to the Monte Collina Bore in the heart of the Cobblers.

Unlike several other outback boredrains that start with the letter 'M', Monte Collina is hardly a hot spot for birds. While a brief visit to Muloorina, Mungerannie or Mirra Mitta bores will typically yield over fifty species of birds, Monte Collina usually hosts a couple of dotterels, plovers and the occasional duck. On my first three-day trip to this inundated pit on the side of the Strzelecki Track I did not see or hear a single zebra finch, which is surely an ominous record for a desert waterhole. When the sandstorms died down enough to allow a wander through the hard mounds of eroded sand the hostility of the environment was confirmed. Finches, parrots and pigeons were surprisingly rare around Monte Collina because there was no grass or grass seed for miles. Bleached rabbit bones, dried white dingo turds and the toilet paper of thoughtless travellers made the site even less appealing. Apart from the relatively unpalatable nitre bush and a few stunted wattles, the rabbits had removed virtually all of the vegetation.

Much to my amazement, criss-crossing through the desolate landscape and around our swags in the mornings were the unmistakable tracks of hopping mice. A pre-breakfast check of Katherine's mammal traps, set the previous night after the marauding ravens had departed, confirmed that the tracks belonged to the rare and beautiful dusky hopping mouse. These pale, feeble-looking mice were distinguished from other hopping mice by the tight twirl of thick hairs on their throat. They are apparently able to survive on seeds and insects left over from the flush of growth that follows a decent rain.

A longer walk around the bore and the surrounding areas highlighted that zebra finches were not the only other animal that was strangely absent. Cats and foxes were apparently rare in the region, despite the abundance of rabbits, their preferred prey. By contrast, dingo tracks and turds were everywhere around the trapping site and we frequently saw and heard Australia's national dog; it was no coincidence that there were few cats or foxes around.

Dingoes serve the same role of keeping foxes and feral cats in check in Australia as wolves and coyotes do in America. Dingoes not only compete with, but also kill cats and foxes. Where dingoes are abundant there are usually few feral predators. A drive down the Borefield Road from Lake Eyre to Roxby demonstrates this relationship elegantly.

Dingoes outnumber cats and foxes at Lake Eyre. In this vast environment where none of the predators are controlled by traps, fences, guns or bait, the dingo rules the roost. As you drive south and into the area that is baited to control dingoes the number of cats escalates. On the southern half of the Borefield Road local hunting enthusiasts have a competition to see how many feral predators they can shoot. Ken Lamb, the hyperactive transport entrepreneur who spends more time organising charity shows and culling ferals than running his business, invariably wins these competitions. Lamby rarely sees any dingoes and shoots about ten cats to every fox. Foxes also take the baits laid for dingoes, leaving the cats free reign in this baited zone. Further south inside the Dog Fence around Roxby where baits are rarely used, local kangaroo shooters typically see more foxes than cats. In the absence of both dingoes and baits, foxes become the dominant predator.

This changing of the guard from one predator to another has been noticed in many places throughout Australia. Parma wallabies in New South Wales and the last remaining bilbies in Queensland persist in areas where high dingo numbers suppress fox densities. An increase in the baiting of dingoes could disturb this balance to the peril of these rare mammals. A further shift down the predator hierarchy can also spell bad news for native wildlife. Researchers in the Shark Bay region of Western Australia have noticed that small native animals have actually suffered in fox-baited areas because of the resultant increase in cat numbers. Closer to home, those islands in Lake Eyre South were probably crawling with mice and painted dragons partly because the dingoes prevented cats from colonising.

These examples suggest that because small mammals are one of the favoured prey of cats, these mammals may actually benefit from high dingo numbers. Monte Collina, with its permanent water, abundance of rabbits and absence of dingo control was a perfect place for dingoes. In turn the abundance of dingoes may have created an enclave for the hopping mice to evade predation by cats or foxes. These little speedsters are usually too fast for a large dingo but are no match for the expert-mousing feral predators.

Cats and foxes are not the only feral animals that are unlikely to be associated with dingoes. Roughly half of the hilly regions in inland Australia are infested with goats, and a fence prevents them from accessing the other hills. This fence stops dingoes that would rapidly wipe out any goats that transgressed into their zone. The section of the Willouran Ranges between Lake Eyre and Lake Torrens that lies inside the Dog Fence is infested with goats. The part outside—in dog country—is goat-less. Goats are a serious problem in southern ranges like the Flinders and Gawler's, yet are absent from the Denison, Musgrave and MacDonnell ranges in dingo country. It is probably no coincidence that the places where dusky hopping mice are still found are outside the Dog Fence.

The Dog Fence was not built to allow goats to prosper in the southern regions or to protect hopping mice in the north. The longest man-made structure in the world was completed in 1946 to enable sheep to graze without predation by Australia's largest terrestrial predator. Ignoring its ecological ramifications, the Dog Fence is a magnificent tribute to our pioneers. Hundreds of thousands of mulga, pine and blackoak trees were hand cut to make posts. These posts were erected into holes chiselled by hand into the rocky terrain. Despite our huge improvements in technology and transport there is no way that we could build another fence of its type today. In today's dollars, construction of the 5,400-kilometre fence would cost in excess of $40,000,000 and securing the necessary money, environmental and heritage approvals would be a nightmare.

Fortuitously for both our pioneers and the modern teams that are responsible for maintaining the Dog Fence, dingoes are no ordinary dogs. Their characteristic howl alludes to their close affinity with wolves, of which they are a sub-species. Unlike many dogs that are agile jumpers, dingoes seldom jump and therefore the fence does not have to be high. Kelpies and collies would treat the five-foot high fence with contempt. Kangaroos, emus and camels take a different path and bash holes through the fence, while rabbits and wombats burrow under and gushing water opens up gaps for dingoes to walk through. As a result, the great fence is reinforced with a thirty-kilometre wide baited zone on the outside to reduce the likelihood of dingoes penetrating the fence. It is again no coincidence that this baited zone includes the same section of the Borefield Road where Lamby shoots so many cats.

Even without the removal of dingoes the Dog Fence affects other wildlife species. Although echidnas are one of Australia's most widespread mammals, they are scarce in most arid areas. Few pastoralists can boast seeing more than a couple of these unique egg-laying mammals on their stations in a lifetime. Echidnas are one of the least agile of our native mammals and are easily stopped by the Dog Fence. The densest population of echidnas that I ever encountered locally was on a field trip to the Dog Fence in the Stuart Range, to the south of William Creek; there we recorded three individuals in three days. Kevie and Shane Oldfield had also never seen an echidna on Clayton Station until two were recently electrocuted in a new section of the Dog Fence. With the exception of echidnas, the few unlucky sleepy lizards that get stuck or the birds that collide with the Dog Fence, its primary legacy is the exclusion of dingoes from a third of the continent.

This elimination of dingoes can have a vast range of environmental and economic implications. In some areas kangaroo numbers have been 100-times greater and emu numbers over twenty-times higher inside the Dog Fence than outside. In the same way that high numbers of domestic stock, rabbits, goats or other feral animals can have serious consequences for the land, even gengelopes have an impact when their numbers are not managed. The innocuous-looking Dog Fence is definitely more than meets the eye.

Despite the widespread acceptance of baiting outside the fence, it is ironic that this approach may actually exacerbate the dingo control problem for the Dog Fence Board. Surprisingly, killing some dingoes may allow others to breed up to challenge both the fence and the doggers who have the job of keeping them out. Laurie Corbett, who has studied dingoes with the CSIRO for many years, has concluded that maintaining the tightly regulated pack structure of dingoes is one of the most effective ways of limiting their population. The dominant female in a pack will often kill the pups of subordinate females and thus keep the population in check. If the dominant female takes a bait the pack may produce more pups, some of which will ultimately disperse into new areas and possibly towards the Dog Fence.

Peter Bird is South Australia's authority on all things dingo and recounts a chilling experience he had with a pack of dingoes in 1990. He was studying the responses of a dingo pack to different fence designs near Jackboot Bore, where I had my plane crash. As Peter watched silently from only thirty metres away, the pack set upon an interloping dingo with a frenzied attack. Minutes later when the yelping had subsided but his heart was still racing, Peter was able to make out the grizzly remains of the gatecrasher that had been ripped apart by the resident pack. Had the dominant individuals of the pack been removed by poison or bullet, the outsider would probably have survived. By suppressing the numbers of young dingoes and vagrant hybrids the naturally regulated pack structure may actually reduce the numbers of calves taken by inexperienced dogs.

Given that the Dog Fence and the associated baited buffer zone to the north has a range of implications beyond keeping the dogs from the sheep, maybe it is time to reconsider the relentless battle that we have waged on dingoes throughout much of the outback. With the downturn in sheep and wool markets more pastoralists inside the fence are turning to cattle, where dingo control is far less imperative. A cattle herd in their prime is not usually troubled by moderate numbers of dingoes. Only weak calves or drought-affected cattle are typically killed. Greg Campbell is the manager for the Kidman properties and estimates the losses to dingoes to be less than one in twenty calves. Some pastoralists concede that by preying on rabbits, kangaroos, pigs or goats, the advantages of dingoes outweigh the losses of a few weak cattle. Maybe instead of disrupting dingo societies and allowing cats to prosper, we should change our tactics and only bait for those dingoes that transgress the fence and penetrate into key areas of the 'inside' country? In those extreme circumstances outside the fence where dingo control can be defended, shouldn't we also insist that land managers target the cats, foxes, goats or pigs that are likely to proliferate to the detriment of the natural ecosystems?

Dingo control is an emotive topic that has seen public opinion swinging back in favour of the dingo. Who could have imagined fifty years ago that most Australians would have been more concerned about a handful of dingoes being shot at Uluru, than the suspicious disappearance of a baby? Perhaps those haunting images of the last thylacine pacing up and down in the Hobart Zoo have reminded us of the disastrous consequences of 'controlling' large Australian predators. Outback travellers who thrill at the sounds of their howls on daybreak, keenly seek out dingoes. Dogs are discouraged from interbreeding with dingoes to maintain, where possible, their true genetic strains. Dingoes are perceived as being quintessentially Australian and icons of the desert. However, in keeping with their paradoxical status even this is open for debate.

Dingoes are relatively recent arrivals to our continent, probably introduced by Asian seafarers about 4,000 years ago. The dingo doubtlessly had a major impact on the wildlife of their new country and led to the extinction of several native species, just like the fox and cat have done in recent years. Dingo numbers built up with increases in the numbers of rabbits, peaking between 1930 and 1950 when water was more widely available from pastoral bores. Laurie Corbett, CSIRO's dingo guru, attributes dingo predation as one of the major causes of the decline of the *Caloprymnus*, those desert rat kangaroos that I hoped to find on the Lake Eyre South islands.

Does their relatively recent arrival and past impact on native species mean that along with cats and foxes, dingoes too should be targeted for extermination? When does an introduced species earn the right to be considered a native? Maybe the dusky hopping mouse, the parma wallaby and the plants persecuted by goats or extreme kangaroo populations inside the fence should have a vote on these issues. Although the situation is far from clear-cut it seems there are some compelling environmental arguments for maintaining dingo populations in some areas where they are currently controlled.

It seems odd that on one side of the fence we take measures to protect a key component of the ecosystem and the genetic integrity of our antipodean wolf, while on the other side we persecute them as vermin. Just as pastoralists and miners must now consider the ecological ramifications of installing new water points and extracting water from the Great Artesian Basin, our better understanding of the ecological effects of dingo control suggests that we reconsider this in some regions. The days when decisions concerning dingo control have been the sole domain of pastoralists are numbered. While shifting the entire fence to a more environmentally beneficial alignment will clearly not be feasible, changing the fuzzy zone and isolated pockets that are relentlessly baited outside the fence is simple.

Dingoes have been the number one enemy for pastoralists and farmers since sheep were first brought to this country. Sheep graziers who have witnessed their sheep mauled by dingoes are understandably reluctant to condone any relaxing of the existing dingo controls. It will be as difficult to change their impressions of the dingo as it will be to convince a one-eyed Carlton supporter to follow Collingwood, or to encourage David Bellamy to start cutting down rainforest trees.

But with changing priorities, new information and improved technologies we should change our management of the dingo in the same way as we are slowly improving the methods that we generate electricity, and the ways that we manage mining, pastoralism and tourism in the outback. Our Australian wolf could well be one of our most valuable allies in protecting unlikely sanctuaries like the windswept Cobbler Sandhills for endangered plants and animals.

25
Tracks in the sand

Something was wrong. While conducting an annual bird survey along a sand dune near Roxby I had a strange sensation that I had lost the plot. The familiar sights, sounds and smells around Roxby seemed almost alien. On this particular morning in the spring of 1996 I felt like a stranger in my own home.

When walking around recording birds it is sometimes possible to let your concentration wander and fail to detect the quiet call of a hidden bird or miss a raptor soaring silently overhead. A week earlier I had been dunked in the 'Devil's Toilet Bowl', one of the most notorious rapids on the Zambezi River, where I spent far too long sucking in mouthfuls of pea-green water. Now as I trudged along the dune I realised that I was coming down with a lurgy from my exposure to Africa's toilet. My legs ached and my dizzy head failed to register anything but the most blatant details. To snap out of this feverish haze I strained my ears and eyes to detect birds in the hope that I would get back into the groove. It didn't work.

Maybe I had lost my ear for the local species or maybe the familiar Aussie bush was not as stimulating as the riot of species I had been treated to in Africa. However, my notepad indicated that I was not doing too badly. I had ticked off the common wrens, woodswallows and other 'regulars' and even recorded several white-fronted honeyeaters that were unusually abundant this year. But no matter how hard I tried I wasn't able to convince myself that I was doing a reasonable job of bird counting.

Suddenly it dawned on me. The marker post at the start of the bird transect had the same number as I had expected, and the patch of mulga trees in the middle of the site also appeared the same as I remembered. It was the sand dunes that were different. The red sand was lightly rippled on the crests of the dunes and it bore the tracks of a myriad of lizards, mice, birds and anonymous nocturnal beetles. In the ten years that I had walked these dunes I had never seen these tracks so clearly. Normally the sand was riddled with the signs of rabbits. Their distinctive scratchings and tracks that were as much a part of the landscape as the dunes themselves were missing.

For the first time in my life I was fixated by searching for rabbit tracks on the dunes. After twenty minutes or so, I finally found the tracks of a lone rabbit. Previously it had been virtually impossible to follow a rabbit's path through the plethora of overlapping tracks. By contrast these tracks stood out like the white line down the middle of a bitumen road. Something serious had happened to the rabbits on this dune. During the ensuing days I found similar results on other dunes in the district. It appeared as if the rabbit calicivirus disease (RCD) had reached Roxby.

RCD, or Rabbit Haemorrhagic Disease as it is correctly known, was first detected in Chinese rabbit farms in 1984. Within a couple of years this virus was responsible for the deaths of over 30,000,000 rabbits in Italy alone. By 1991, the virus was imported to Australia under quarantine to test whether our native animals were susceptible. A few years later, after failing to infect any native species, a field trial of the virus was approved for Wardang Island, off the coast of the Yorke Peninsula between Adelaide and Roxby. Incidentally this is the same location that the other great hope for rabbit control in Australia—myxomatosis—was trialed in 1937. In late September 1995, soon after the field trial had started, RCD 'escaped' from Wardang Island to the mainland, devastating rabbit populations and arousing huge public interest.

I asked Zoë Bowen, who had graduated from the knobtail searches to inherit my rabbit-monitoring role, when the calicivirus had reached Roxby. She was not aware that it had. Likewise, none of the neighbouring pastoralists had witnessed the signs that RCD was here, like the dead rabbits on stinking warrens that had first attracted media interest when RCD obliterated rabbits in the Flinders Ranges. The only way to tell that RCD was responsible for the dramatic decline in rabbits was to extract the blood from twenty rabbits and test for antibodies to the virus. Rabbits less than six weeks of age are largely resistant to RCD and hence generally survive an outbreak. These survivors serve as markers that the virus has reached an area.

Rabbit densities had been monitored for the past decade to help explain fluctuations in vegetation growth at mine rehabilitation sites. Every two months, along with counting rabbits along set transects, twenty rabbits were shot to record their weight and reproductive condition. The age of the rabbits was determined by drying and weighing their eye lens, which continue to grow with age like those of humans. Typically only a couple of hours were needed to fill an esky with bunnies. Now, after RCD, the rabbit hunt took several nights.

Eventually blood was collected from enough rabbits to confirm that RCD had arrived. However, determining where it had struck was more difficult. Robby Savage told me that he could identify patches of nearby Mulgaria Station where the virus had been extremely successful and other areas where it seemed to have missed out. It appeared that RCD was operating like a bushfire, leaping ahead along some fronts and leaving unburnt pockets behind. The fuel for RCD was rabbits and it burnt most effectively where rabbit populations were high. Rather than oxygen and wind, flies appeared to be the agent responsible for transmitting the RCD fire. RCD could also occur repeatedly: once rabbit populations had returned to a critical density in certain areas the calicivirus could strike again. This could create some areas with virtually no rabbits and others with gradually increasing numbers.

RCD provided the new great hope for the successful control of rabbits since they arrived in 1888. With powerful kidneys and an ability to tolerate losing a third of their body weight, rabbits are surprisingly well suited to living in arid regions. The rabbit, which first had to acclimatise to the cold English climate after being introduced from Spain in the eleventh century, was now demonstrating its survival skills in an even harsher environment. They grew longer ears and lightened the pigment in their fur to dissipate and reflect heat more efficiently than their ancestors. Most rabbits that live in pale sandy deserts now exhibit the normally rare recessive 'yellow' gene that accounts for their ginger colouration. Within a few decades the rabbit had firmly established itself as the major scourge of inland Australia, a position that it held until the outbreak of RCD.

Just how bad rabbits have been at times is difficult to comprehend. Keith Greenfield can remember trapping up to 2,000 rabbits a day in a single trap yard built on Billa kalina Station. That is 2,000 rabbits in an area half the size of a tennis court! He also remembers hordes of rabbits accumulating along netting fences in such numbers that late arrivals could walk over the fences on the piles of corpses. Similarly Brian Cooke, Australia's pre-eminent rabbit control researcher, estimated that rabbits reach densities of 3,500 per square kilometre in plague times. Since rabbit densities as low as five per square kilometre are still too high to allow the regeneration of some trees and shrubs, the impact of these rabbit plagues are devastating.

Like Ferguson's cattle around the Bubbler, these vast hordes of rabbits eat out every palatable morsel, exposing the desert soils and sands to unprecedented erosion. With the exception of a few potential beneficiaries such as the hopping mice at Monte Collina Bore, most native mammals that could eek out a living in such degraded landscapes fell victim to the huge populations of feral and native predators that thrived on the rabbit plagues. Burrowing bettongs or boodies, which helped the spread of rabbits throughout much of inland Australia, were rapidly displaced from their palatial burrows by the faster breeding and more predator-savvy rabbits.

Because of their infamous breeding credentials and their ability to adapt to different environments, rabbits are exceedingly difficult to control. Around the town of Roxby tens of thousands of dollars were spent on rabbit control in the late 1980s and early 1990s to enable the establishment of gardens, the golf course and oval. Despite fencing and consistent poisoning, shooting and the fumigation of warrens, the rabbits maintained the upper hand for years. We knew there had to be a better way—a biological solution to controlling rabbits. Brian Cooke was our best bet.

WMC sponsored this quietly spoken biologist in his studies of the Spanish rabbit flea. Brian hoped that the flea would prove to be an effective vector for myxomatosis in our dry region like it was in the dry parts of Spain. Fleas could help spread myxomatosis to reduce rabbit numbers in arid Australia like mosquitoes do in wetter areas. Because of our long-term monitoring of rabbit populations, Roxby was selected as a principal arid zone trial site for these fleas. In March 1993 Brian brought his prized loot of 5,500 fleas which we released into six rabbit warrens. Just to make sure that we had a large enough seed population of the fleas, Brian released a further 6,000 in other warrens three months later.

Before releasing these additional fleas we scoured the entrances of the original six warrens for evidence that the initial release had been successful. No fleas were found. Suspecting, or rather hoping, that the fleas were either attached to rabbits or were deeper in the warrens we went out that night to sample active rabbits. After shooting rabbits near the target warrens, we sprinted over and placed them in plastic bags before any attendant fleas had time to jump from their sinking ship. A fine comb was then used to dislodge any fleas from the rabbit fur.

Despite the thousands of fleas that we had released we were only able to relocate one Spanish rabbit flea through this technique. However, another turned up a few hours later in the Environmental Laboratory as Brian was fielding a barrage of questions about fleas. Our secretary was particularly concerned that the rabbit fleas would be attracted to people. Right on cue, Brian felt movement through his tangly beard. A flea must have jumped on board when he had his head down a warren. As nonchalantly as possible, Brian jammed his finger into his beard, pinning the flea to his chin while he casually sauntered out of the room.

In February 1994 we released even more fleas, but this time they were infected with the myxomatosis virus. There were plenty of rabbits around and if our two previous releases had been successful there should have been enough fleas to spread the virus. With expectation and apprehension we closely monitored the rabbits in our flea-release area for the telltale signs of a myxo outbreak. Days led to weeks, which led to months of watching perfectly healthy bunnies chewing away at everything in sight.

Although Brian had recorded some encouraging increases in myxo following the release of fleas elsewhere, their effect at Roxby was conspicuously unspectacular. Nevertheless it was while studying the fleas in Spain that Brian observed the initial effects of RCD that ultimately led to the RCD trials in Australia. Furthermore, we now know that the rabbit fleas can transmit RCD and hence they may be one of the silent conspirators in the ongoing war against rabbits. Despite all of the money and effort we had put into rabbit control at Roxby, RCD had provided unimaginable results despite no local effort. As a result, in the late 1990s we witnessed a short-lived recovery in the outback environment that was similar to that which occurred after myxo did its thing in the early 1950s. It was an exciting and invigorating time.

The most immediate response to the reduction in rabbits was evident in the grasses. Previously, within weeks or months of luxuriant rain-induced growth, most of the grass in rabbit country had been mowed down like a bowling green. Now the grasses remained tall, even after they had dried off, and provided cover for animals and a food bonanza for termites and crickets. Other plants responded in a similarly spectacular manner.

Parakeelya is a short-lived succulent plant that is a valuable food and water source for a range of animals including domestic stock and rabbits. Aboriginal people have eaten its crisp, slightly acidic-tasting leaves for centuries and I too enjoy nibbling on the finger-like leaves while wandering around the dunes in spring. Parakeelya is also blessed with vivid magenta flowers. These blooms open for a few hours around midday, thus earning it the common names of 'one o'clock' or 'lunchtime' flower. In most years, scattered clumps of these spectacular flowers highlighted the areas that received autumn or winter rains. However, outback communities were enthralled in the spring of 1996 when every day for a month or so the one o'clock flowers turned the orange dunes a psychedelic purple. Although the experiment with the pines indicated that rabbits were not always the prime culprits in preventing plant recruitment, the parakeelya was a vivid reminder that high rabbit numbers had a devastating effect on the vegetation, soil stability and functioning of ecosystems.

One of the first animals to respond to a life without rabbits was the hopping mouse. In 1997 Katherine was convinced that she had seen hopping mouse tracks on a sand dune at Roxby, despite these mice never being recorded locally. For days we followed tracks and set traps, eager to find out which species had excursioned down from their normal northern haunts. Eventually we caught a spinifex hopping mouse and within weeks saw plenty of signs of them on nearby dunes. Their distinctive tracks and popholes soon turned up on nearly every dune and we were often rewarded with these long-legged mice with their elegant feathery tails in our traps. Presumably the hopping mice, along with other rodents, were responding to both the increase in grass seeds or the decrease in predators following the demise of rabbits.

The plains rats, those rare 'guinea pigs' that were the serendipitous find of my inland taipan searches, also suddenly appeared. The reduction in rabbits following RCD was just the impetus they needed. Kelli-Jo Lamb, who had inherited my job as 'The Ecologist' in 1998, quickly made her mark by capturing a couple of plains rats from sites where they had not been recorded since trapping had commenced a decade earlier.

The response to the reduction in rabbits was certainly exciting, but at the same time it was a little frightening. Rabbits are bad news in Australia but no-one was sure of the implications of the sudden demise of one of the most abundant and influential outback species. American ecologists have names for types of organisms that make up an ecosystem. Due to their important and varied roles the Yanks could well consider rabbits to be a 'keystone species' throughout much of central Australia. Reduction in rabbit numbers may have profound effects on many other animals and plants that are influenced by them, and not all of these effects are positive. The removal of rabbits actually spells hardship for several native animals.

Rabbits are accomplished diggers that can construct large warrens with up to thirty or more entrances. Through their digging rabbits are responsible for considerable and not always desirable soil turnover and this creation of holes, divots and mounds all effect water penetration and run-off. As well as being a keystone species the Yanks would also call rabbits 'ecosystem engineers' for fairly obvious reasons.

Rabbit burrows are refuges for many animals. Their role parallels that of the prairie dog in North American rangelands. At least eleven vertebrate species depend upon prairie dog burrows for shelter. In the Australian arid zone echidnas, native rats and mice, snakes, lizards, and creepy-crawlies use rabbit burrows extensively, particularly in hard ground where burrowing is difficult. Several other of our burrow-using species could not persist in their present numbers or range of habitats were it not for the presence of rabbit burrows. Even birds use rabbit warrens. I have seen young inland dotterels retreating down rabbit holes when threatened. Other birds such as red-backed kingfishers, white-backed swallows and rainbow bee-eaters may construct their nesting burrows in large rabbit warrens if earth banks or cliffs are not available.

Rabbits are also important prey for large native predators. Perenties, woma pythons, dingoes and a series of raptors all dine regularly on rabbits. When rabbits were thick on the ground prior to

RCD I typically found four or five huge wedgey nests in native pine trees within the Olympic Dam lease. A quick look among the fallen sticks under a wedgey nest always revealed many bleached-white rabbit skulls. It therefore did not come as much of a surprise that wedgies didn't breed locally for four years after the rabbit numbers crashed. This was not cause for urgent concern because wedgies live for such a long time, but continued breeding problems for rabbit-eating raptors could have huge impacts. In the UK buzzard populations halved in the late 1950s when rabbit numbers crashed. A similar fate could face several of our own eagle, kite, hawk and buzzard species.

The demise of rabbits not only challenges the eating and sheltering habits of native species. The removal of one of the dominant herbivores from many arid environments will inevitably lead to increased vegetation cover. Greater fuel loads will in turn increase the frequency and intensity of fires in some habitats, with their own unique affects on the rest of the ecosystem. It hurts to admit, but those destructive little buggers actually perform some useful services to the outback environment.

With rabbits featuring so prominently in the ecology of native species it is difficult to establish the pre-rabbit scenario for these species. Rabbits probably replaced the role of native burrowing mammals such as the bilby and the boodie. Brush-tailed possums used to live in boodie burrows in cental Australia and subsequently used rabbit burrows. The extensive inventory of native mammals that have become locally or totally extinct since the arrival of rabbits, also probably lowered fire fuel loads and supported populations of large predators, although maybe not to the same levels as with rabbits.

Ideally if rabbits are to be eliminated they should be replaced with the full suite of native mammals that were displaced a century ago. These mammals could then provide burrows, food and other ecosystem services and the rabbits won't be missed at all. Of course this is not possible because many of our native species are now extinct. Most of those that are still hanging on are in such low numbers that they will not be able to replace the role of rabbits for many decades. Conservation of boodies and bilbies is therefore far more important than the preservation of cute and cuddly desert icons. Hopefully, just as their chocolate effigies are now gradually replacing the outdated and inappropriate Easter bunny, our native diggers will become the outback's future ecosystem engineers, just like they were in the past.

Before we get too carried away with visions of a rabbit-free Australia we have to take stock of the situation. With the exception of a few intensively managed areas, rabbits have not gone. There will probably always be rabbits in Australia. Like myxomatosis, RCD will not eliminate rabbits and rabbit populations are already rebounding like they did after myxo's heyday. Sandswimmers, barking geckos and native mice are now able to make the most of their refurbished palatial burrows, but the clean sand that in 1996 was only inscribed with indigenous tracks is once again being tarnished by feral graffiti.

Our biggest concern should not be what happens to our environment if rabbits disappear altogether, but what happens if they boom again. The war on rabbits must continue to be fought with gusto, both in the research laboratories and in the bush. If we don't maintain control of these little blighters, then there is little point in even dreaming of an outback refurbished with viable populations of native keystone species and ecosystem engineers.

26
Cobwebs in crevices

We were in the best place in the world to observe a unique population of black-footed rock wallabies. The waterhole where they would drink was below us and the caves, cracks and boulder piles where they sheltered during the day were clearly visible on the opposite cliff. Not only was this the perfect spot to count the wallabies, but this snug round cave perched halfway up the cliff was an awesome place to camp. Katherine and I felt privileged and excited. It had taken an exhausting twenty-minute climb to haul our swag up the steep cliff. Katherine had started her first watch as the dark purple shadows of sunset gradually engulfed the eastern cliff. Meanwhile I clambered back down to the dry creek bed to collect a bottle of water and some raisin bread that was to be our dinner and breakfast.

As I started back up the cliff I was distracted by a flock of little woodswallows being sucked under a small overhang like iron filings onto a magnet. Within a few seconds twenty of these birds resembling sleek dark paper jets had crowded headfirst into a tiny fissure, their tails fanning outwards like a giant feathery starfish. Below us a pair of corellas screeched sweet-nothings to each other as they meandered up the eucalypt-lined creek from the flat gibber plains. A peregrine falcon raucously announced that she too was roosting for the night high on the red cliff. This place was so special and the *pièce de résistance* was yet to emerge.

Katherine ushered me back into the cave quietly without looking up from her binoculars. 'No', she said in a hushed voice, 'not yet'. By now the last rays of sunshine were just lighting up the crest of the opposite cliff. I pulled out the infrared night scope and trained it on a large crevice. Earlier I had clambered there and was awe-struck by the deep covering of wallaby turds on the rock floor. Now my eyes adjusted to the haze of deep green shadows and bright green stars that indicated lighter-coloured rocks or leaves through the night scope. It was not until I had seen a kestrel flying to its roost that I was convinced that I would be able to distinguish a wallaby in this strange green light. For half an hour we took turns looking through the scope, reluctantly handing it over when our eyes went hazy.

There was no moon and the gully rapidly became too dark to use the scope unaided so we pulled out the infrared spotlight. Lining up the spotlight and the scope was far more difficult than either of us had anticipated. It is not easy to follow the illuminated zone of a hand-held spotlight with binoculars. It was even more difficult for one of us to scan the black cliffs through the night scope while the other was slowly 'painting' the cliffs with 'invisible' infrared light. Eventually we decided that we were probably better off listening for the wallabies so we lay down in our cave, occasionally pulling out the spotlight to watch a euro scramble down the cliff or to see a nightjar fluttering like a huge moth down the valley.

Katherine and I were worried, really worried. A few months earlier we had seen the wallabies in this gully a number of times. Phil Gee and his wife Ifeta, who had studied zoology with me at university, had seen them dozens of times since they discovered this isolated Davenport Range population in 1990. In the past we had not holed up in caves to locate them. Typically a few of the small grey wallabies with dark feet and flanks could be seen bounding up the cliffs at any time of day. These glimpses of fleeing wallabies made it difficult to determine the size or viability of the population. That was why we were sitting quietly in the cave.

When Phil and Ifeta, along with wallaby researcher Mark Eldridge, had trapped six individuals seven years earlier in 1992, they estimated that the colony consisted of approximately thirty individuals. Blood samples taken from the trapped wallabies, which were immediately released, indicated that these wallabies were the same species of black-footed rock wallabies that lived in several other rocky ranges in central Australia. Impaja TV commercials repeatedly reminded us that if we camped at Heavitree Gap at Alice Springs we could feed and photograph the same species. Previous measurements of the animals we were hoping to count indicated that their ears were smaller and that their genetics were different from other populations. Not only were they a genetically distinct population but they were found at least 400 kilometres south-east of any other populations of black-footed rock wallabies that had survived. Here they were largely restricted to the gorge where we were camped, although some were occasionally found in the neighbouring, less suitable gorges.

Like the *Caloprymnus* and a host of other medium-sized mammals, black-footed rock wallabies were in that critical weight range of mammals that was most affected by predation by foxes and cats. Rock wallabies used to be considerably more widespread than they are now, with either the black-footed or their spunky cousins, the yellow-footed rock wallaby, inhabiting most of the rocky ranges in northern South Australia. Wallabies used to live in the Willouran Ranges just south of Lake Eyre and even in rocky gullies on the western side of Lake Torrens. Now they were gone or gradually disappearing, even from their stronghold in the Anangu-Pitjantjatjara Lands in the far north-west of the State. Katherine had drawn up a plan to study this isolated population a few hours drive north of William Creek. Our aim was to assess whether the wallaby numbers were stable and to suggest whether any active management was necessary to safeguard them.

Two afternoons ago we had driven into the ranges following Phil on his motorbike. After we had left the old bullocky track that was the precursor of the overland telegraph line we endured a gruelling, bone-jarring cross-country drive in my troop carrier. The troopy took a battering—I lost the towbar electrical connections and dented several other accessories and sump guards while crashing through steep washaways and negotiating boulder-strewn creeks. At first light the following morning we emerged from our swags and entered the main gorge scoffing some breakfast on the move. Poor old Phil couldn't believe that we would not ease into the day with a billy cuppa first. However, neither Katherine or I drink tea or coffee or like dawdling around, especially when there are exciting places to explore.

The wallaby gorge was a couple of kilometres long, and with the exception of a 500-metre section between two steep rocky slopes, the dry creek bed was littered with donkey dung. Phil reckoned that the steep section was impenetrable to donkeys and that this was one of the main reasons that the wallabies had managed to survive. Donkeys drank or polluted most of the waters in the ranges but were unable to access a near-permanent rockhole in the steep section. Although we had found a couple of wallaby turds by the rockhole we did not see any wallabies that day. Nor did we see wallabies in the gorge when we snuck through at dawn and dusk the following day. That was why we decided to camp in the cave—to let the wallabies come to us.

It was not only concern for the wallabies that interrupted our sleep in the little round cave. Caves look like idyllic places to camp but the floor of this one was sloping and uneven and the musty smell of wallaby and euro turds irritated my nose. It was with relief that the sky began to lighten and we could abandon thoughts of sleep and resume our searching. For nearly two hours after dawn we lay motionless with just our binoculars peering from the cave, hoping to count the wallabies as they emerged to sun themselves on the rocks. We saw none.

Later that day, like the previous days, we walked transects along compass bearings for several kilometres counting macropods (red kangaroos, euros and wallabies), predators (dingoes, foxes and cats), competitors (donkeys and rabbits) and the dung, tracks or warrens of all these species. We deviated from our transects to check out any nearby caves, overhangs or crevices where the wallabies could hole up. In 'bat cave' gorge, named by Phil and Ifeta, we climbed up to a small cave from which the gorge takes its name. The handful of old grey wallaby turds at the entrance indicated that the wallabies had visited the cave in the past, but the absence of black or glossy turds suggested that they had moved on.

We entered the cave as quietly as we could and shone our torches on the roof. The discreet liquid-paper X on the inside of the cave indicated, like most caves in the ranges, that Phil had already been here. Six large bats with prominent spiky tails curled over their backs like gryllid crickets peered back at us, their eyes glowing in the torchlight. The rare hills sheath-tailed bat was one of the most significant animals that still called the Davenports home. Other important findings in the area include southerly records of the fat-tailed antechinus and long-snouted tree dragons as well as populations of the rarely seen desert mouse, slender blue-tongue, echidna and painted firetail finches. As far as we knew these other rare species were hanging on ok, but the situation for the wallabies appeared dire.

At the end of our long weekend we left the ranges dispirited and worried. Not seeing any wallabies did not mean that they had all gone. After all, they had presumably lived in these ranges for thousands of years and were unlikely to disappear within eight years of their discovery. On our next available weekend we headed back. Phil could not make it but we were fortunate that our tyre tracks were still visible and with considerable relief we found our way back to our campsite outside the wallaby gorge. This time we were not going to rely on transects or infrared scopes, we were armed with a much more powerful tool: peanut-paste sandwiches.

We poked sandwiches into over sixty caves and crevices. Most of these were in places inaccessible to euros, donkeys and other potential sandwich bandits. Each place we dropped a sandwich was marked by a strip of bright pink flagging tape so that we could go back later and check whether they had been taken or not.

On the way back from distributing the sandwiches Katherine found a wallaby carcass in a crevice just below the round cave where we had camped on our previous visit. When I responded to her agitated call I found her looking despondent. Dangling from her hand was a dehydrated wallaby body, intact except for one claw. Not far away from her grizzly discovery I found two wallaby jawbones within a metre of a fox turd. We had still not seen a live wallaby. In fact none had been seen since a National Parks reccie a couple of months earlier. The equivalent survey the previous year yielded a dozen individuals. Despite climbing into all of the likely crevices several times we had only found fifteen black turds and even these could have been weeks or months old. The caves were worryingly quiet. Even more ominous than the lack of wallaby sightings and the scarcity of fresh turds were the thick shrouds of cobwebs across the crevices that used to be their homes.

Two days later we revisited all of the baits. None had been taken. Because we did not want to attract any predators into the key wallaby habitat we retrieved all of the dried out sandwiches. This was not an easy task as we had sometimes thrown them deep into a crevice that was now either difficult to reach, or nearly impossible to get into. It had started raining and the rocks were slippery as we edged along crevices and crumbly cliffs. I can remember contemplating that this was much more dangerous than looking for taipans, especially since we were isolated from the relative security of William Creek by nearly four hours of driving. The drive would take much longer or could have been totally out of the question if the rain had really set in.

We were now convinced that the wallabies were no longer in their main gorge. But others we spoke to were less convinced. 'They would still be there,' we were told. Apparently because they are so secretive and well camouflaged, rock wallabies are often overlooked on one or more searches even though they are subsequently relocated. 'They couldn't just disappear overnight, maybe they just moved?' people proposed. Maybe they had.

A few weeks later Katherine and I banged and crashed our way back to the Davenports once more. We decided to increase our search area and had now received permission from the National Parks mob to distribute some meat baits for foxes and dingoes. I felt guilty poisoning dingoes on their home turf but reconciled that the loss of a few dingoes, along with some foxes, was our only chance to save the wallabies. Several researchers had shown that wallaby populations increased in areas where foxes were baited, but declined or even became extinct where they were not managed.

A couple of mates from Roxby, Jackie and Keith, had come along to give us a hand because we were trying to cover twenty square kilometres as comprehensively as possible. Phil was also available for this trip and he drew up maps for us to follow through the rugged terrain, while he used his motorbike to bait the flatter areas and to check some outlying rock outcrops. One of these sites was a series of large boulders on top of a hill where a Kidman's worker had recently seen some promising-looking scats. We set off in different directions, each hopeful that we would be the one to locate where the wallabies had moved to.

For hours we each walked through the timeless ranges with no sign of the twentieth century to distract us. No car tracks, rubbish or sounds other than those that could have been heard by the people who had engraved figurines into the rocks at the nearby springs generations ago.

After climbing down the rugged spinifex-clad Mt Jarvis my route took me through the shaly foothills of the ranges. Every 200 metres I scratched a divot with my boot and buried a meat bait injected with 1080 poison. I was following a small creek and despite climbing up into overhangs and caves found no evidence of wallabies. The creek channelled into a low narrow gorge that led to a series of small waterholes fringed with bulrushes. I had obviously reached a spring and when I found a large round cave next to the biggest waterhole, I figured I was at Edith Spring. I vaguely remembered Bobby Hunter's story about a round cave that he had sometimes called into when he used to muster cattle on The Peake Station.

I thought there may be Aboriginal carvings in this cave adjacent to the water but I couldn't find any; indeed the cave was a lot less spectacular than it looked from the outside. I was just leaving when I noticed a handful of fresh wallaby droppings, right near the entrance. They were black and shiny and unmistakably different from the more spherical and coarser joey euro turds seen in other caves. I pocketed a few scats to confirm my find.

I was stoked! This was the first fresh evidence of wallabies that we had found for several months. The wallabies must have moved down to here. Within minutes I was bound to see a wallaby or find more caves that they were using. I looked around with renewed vigour. There were little caves everywhere. After scrambling into many of them I became increasingly dismayed. No wallabies and no turds, but plenty of cobwebs. The country didn't even look right. The low grey crumbly cliffs were almost devoid of grass, unlike the main gorge that was chockers with good wallaby tucker. The turds at the spring were probably from a dispersing individual that had only stayed for a few hours.

I left the foothills near the springs with mixed feelings. As I headed back home into the setting sun I counted the creeks that I crossed so that I knew which one to take to our campsite at the base of the range. Finally just on dark I smelt the campfire and exhaustedly walked up to the others, hoping that they had better luck in locating the wallabies. They had all drawn blanks. Even the droppings at the boulder spot that Phil had checked were from euros. The following day the news was no better. For all we knew there had been only one wandering wallaby within the past few months. Surely the population was doomed.

Following yet another visit with no evidence of wallabies, the National Parks and Wildlife Service were eventually prompted into action. They decided that *their* 'wallaby team', with experience in counting yellow-footed rock wallabies and baiting for their predators in the Flinders Rangers would get fired up. Previously the National Parks mob had decided that the yellow-footed rock wallaby was a higher priority for their limited management dollars than the slightly smaller black-footed rock wallaby. That decision seemed to make sense. The yellow-footed rock wallaby is one of South Australia's glamour species. It occupies areas that are more spectacular and that attract far greater tourist numbers than the isolated low range of hills where these black-footed rock wallabies lived. National Parks knew that the yellow-footed rock wallaby had suffered huge declines and was now making an encouraging recovery following intensive baiting. Hence, in the absence of proof that the Davenport Ranges black-footed rock wallabies were under any real threat, they committed their sparse funds to the established program.

With the 'wallaby team' in control we were not allowed to head off for the day with a water bottle, handful of muesli bars and a bucket of baits, like we had done previously. Everyone, including the station blokes on bikes who were helping with the baiting outside the ranges, was required to carry detailed maps, UHF radios and adhere to frequent radio calls. In order to drop the walkers at isolated locations and maintain permanent radio contact with everyone, National Parks had also hired a chopper.

Helicopters can be the most fantastic craft for getting around remote places and they can also be the worst. Ever since my first ride a few years earlier when the chopper dropped out of the sky I have always been a bit nervous about getting airborne in a helicopter. Unlike the cool head of the pilot of my plane incident at Lake Eyre, the panicking chopper pilot had screamed to me through my earphones 'What the fuck's going on?' I had no idea what was going on except that I had heard some loud cracks of breaking metal and that the ground was rushing up at us. The memory of these few seconds is indelibly etched into my memory, although unexplainably I cannot remember the actual impact. Even though the chopper was a bit worse for wear we both walked away from the crash. The fact that our escape was lucky was reinforced a few years later when a pilot and his passenger in the very same chopper died when it ditched near Marla.

Despite my dodgy flying experiences it was not the prospect of being airborne in a bubble without wings that concerned us most. Katherine and I found it frustrating how slowly everything was conducted with the team of Government workers. The chopper had broken down soon after arriving. Because it was out of action, the National Parks team could not guarantee half-hourly radio contact would be made with all walkers, and hence it was considered too dangerous to head out far from camp. We were driven by passion; the National Parks team were constrained by regulations. The mining and even the pastoral industries with which I was involved were also increasingly governed by more and more procedures and protocols. These were designed to improve safety and reduce liabilities, which although frustrating, are undeniably positive developments. Nevertheless, Katherine and I preferred it when we could escape the bureaucratic shackles and assume responsibility for our own safety on weekends and holidays.

Towards the end of the second of these National Parks-coordinated trips, Katherine and I met at the top of the wallaby gorge. We did not speak as we slowly tiptoed down the dry creek. We did not have to. Both of us were experiencing one of the saddest hours of our lives as the reality sunk in. The wallabies had gone. This unique population had become extinct within a decade of being discovered. No-one had acted fast enough. Everyone with any knowledge of the wallabies was in some way guilty.

Katherine and I diverged again as we headed down the gorge for the last time. She was having one last look down a side gorge that used to support wallabies; I climbed once again into one of my favourite caves. I could see the rockface where I had watched wallabies scampering a couple of years earlier. Just below was the waterhole where Katherine had found the last carcass a year before. Everything looked all right: the wallaby grass, the big red boulders and the finches zipping down to have a drink. But everything wasn't all right. It was shit. I had visited this crevice three times in the past ten months and each time the wallaby turds on the floor were a bit paler and a bit drier. The cobwebs that I cleared each time had reformed. I could not remember ever feeling so guilty, so sad and so helpless.

What had gone wrong? What had happened to these wallabies that had been isolated from tourists, cattle, mines, pollution and even donkeys? Katherine and I suspected that their population had been restricted to an unviably small level by the high numbers of predators and competitors. In order for them to survive for many centuries the wallaby population was probably once much larger, and occupied a far greater range than when Phil and Ifeta found them in 1990. Foxes alone could have restricted the wallabies to just the most precipitous gorges, and donkeys could well have fouled or drunk the waterholes that were once important for them. Dingo, wedgey and cat populations that were buoyed by high rabbit numbers would have amplified the pressure, restricting the wallabies to only their best-fortified crevices. It is possible that following RCD this bevy of predators turned their attention on the few remaining wallabies. Later, close inspection of predator turds indicated that many rock wallabies had been eaten during this period.

Sitting in that cave it dawned on me that what I had been doing for the past decade was not enough. I had been discovering and researching many of the ecological secrets that the outback held. I had been developing better techniques for monitoring land-use impacts and identifying species that were sensitive to disturbance. But none of these things had helped the wallabies at all. By itself, none of this research was actually saving the critters from the threats that Steve Morton and others had shown us were eroding our biodiversity. I had let down the very environment and critters that were my motivation.

The loudening beat of a chopper broke my misery. 'You ready for a pickup John?' crackled the handheld radio. I clambered up the cliff to the narrow ridgetop where the chopper was touching down lightly. We whirled past the gorge and back over the low hills to our campsite.

Leaving was easy; forgetting is impossible.

27
Who's responsible?

After reading accounts of mammals in the 1930s that have since become extinct, I was frustrated that little had been done to save these creatures. So many excuses could be made. In those days outback residents were so busy trying to survive, that they were probably not aware of the impending demise of these species. Back then no-one had the money, knowledge or support to conserve the critters that lived out in the boondocks. But what about now? For several generations Australians have been aware of our atrocious legacy of mammal extinctions and we have also understood the causes of these problems. I used to think 'If only we knew back then what we know now, how different the arid zone would be.' Or, more fancifully, 'Imagine if I had been around back then and could have helped to save those species'.

What a joke. I was around in exactly the circumstances that I had dreamed of and the result had been no better. I found no solace in the fact that the Davenport Ranges' wallabies were but one variety of black-footed rock wallaby and that other populations of the same species still survived elsewhere. Hugh Possingham's students had confirmed the importance of populations. Varieties are the base from which natural selection upon individuals is made. Even Charles Darwin, who had a pretty good grasp on this subject, concluded that in an evolutionary sense 'there is no fundamental distinction between species and varieties'. Furthermore, in terms of the natural balance between the plants and animals of the Davenport Ranges' ecosystem, it was inconsequential that related wallabies lived elsewhere. The Davenports would forever be poorer and more vulnerable to other changes following the loss of one of its distinctive animals.

Phil and Ifeta and then Katherine and I could have acted earlier than we did to save these wallabies. We all now wish that we had. The Gees in particular felt that the wallabies were best served by not attracting attention that could have been detrimental to both the wallabies and the local pastoralists. Only key scientists at the National Parks and Wildlife Service and the museum along with neighbouring pastoralists were aware of the colony before their plight appeared imminent.

Of course the National Parks and Wildlife Service could have, and in hindsight, should have acted more promptly. If a fox-control program had commenced soon after the wallabies had been found, or at least immediately post-RCD, the wallabies would probably still be in the Davenports today. However we shouldn't have to rely on a few under-resourced bureaucrats to instigate and implement all conservation initiatives. Fortunately many National Parks have volunteer groups who willingly supply labour and resources to assist when the paid staff cannot keep up with the workload. But these wallabies were not in a park, they were in the back paddocks of a couple of pastoral stations.

Pastoralists also have responsibilities to look after wildlife. The dwindling population of wallabies on their leases may have been saved if the local pastoralists had proactively instigated or campaigned for predator control in the Davenports. But the two closest station homesteads were well over an hour's drive from the colony and cattle seldom strayed into the hills. It was only through years of gaining the locals' trust that Phil and Ifeta were allowed into the Davenports, which were partially protected by this restricted access to visitors. Therefore the local pastoralists were neither directly responsible for their demise, nor wholly responsible for saving the wallabies.

At a Roxby bbq soon after one of our ominous wallaby surveys, I was approached independently by two blokes who had never seen the wallabies, nor even heard of the hills where they had been found. One was a pastoralist, the other a transport contractor. Both offered to assist the conservation of the wallabies in any way that they could, and each proposed a $5,000 donation to the cause. Neither of these unlikely philanthropists was grandstanding to an audience, nor had anything untoward to gain.

I was heartened by their concern and surprised and thankful for their offers. Bush communities in Australia are incredibly resourceful and generous with the little spare time and money that they have. Consider the response to natural 'disasters' like flood, drought or fire. Look how many planes have been purchased using outback fundraisers for the Flying Doctor. Maybe if the wallabies had been given more exposure, the enthusiasm and resources of the regional community may have got them through their tough time.

The next time I was drinking with some locals a very different opinion was expressed. A steady flow of beers and rum had opened up the conversation at the William Creek Pub. Katherine was firing on all cylinders and I suspect that Grant McSporran was either playing the devil's advocate or just winding Katherine up. Grant was the manager of nearby Anna Creek Station and had told Katherine that the wallabies would have been ok if we had not gone in and interfered. It was a compelling argument. The wallabies had survived for many thousands of years without anyone meddling in their affairs. All of a sudden within a decade of their 'discovery' they had been wiped out. Maybe this monitoring and interest in the species was the cause of their problems?

Overeager scientists have been held responsible for the extinction of other populations or species before. Within a decade of the discovery of a Queensland frog that raises its tadpoles within its stomach, the frog was extinct. Because scientists had collected the last few known specimens, some commentators blamed over-collecting for the fate of this gastric-brooding animal. Similarly, some elderly Aboriginal people attribute the loss of some mammals to collecting by museums, since the last specimens they saw were collected for, or displayed in museums. Neither of these arguments are valid and Katherine was not having a bar of Grant's assertion that we should have left the wallabies alone. She pointed out that ignoring threatened species or ecosystems is a cop-out. There are no longer any ecosystems that have not been directly or indirectly affected over the past few centuries by the unbalancing effects of mankind. Try to find a locality in Australia that maintains its original plants, animals and soil profile or one that has not been invaded by introduced weeds or animals? Like it or not we have upset the applecart. We can't just sit back and watch the apples roll out. Katherine argued that the wallabies' problem was not that we had interfered, but that we did not act early enough.

There are some heartening examples of last-ditch 'interference' that have probably saved threatened species. A handful of mala were taken from the wild and successfully bred in captivity soon before a fox and a fire wiped out the remaining two mainland populations of these marsupials. Malleefowl and yellow-footed rock-wallabies only survive in their current ranges and numbers because we actively manage their other threats. In most cases where we are aware of an impending crisis we know what needs to be done to save a species, or even whole ecological communities. The fate of our threatened species therefore rests largely with whether the finances or motivation are available to halt their declines.

Unfortunately the glamour species like rock-wallabies and malleefowl that we target for protection are only the tip of the proverbial iceberg. Species, habitats and environments typically only receive attention and protection if they are recognised as being special. This is a classic 'Catch 22' because they are only recognised as being special if they have a story, something on which to hang a newspaper article, a documentary or a conservation drive. Stories can only be told about special critters or places if someone is intrigued, motivated and sufficiently financed and skilled to study them. Animals or regions without a story are just another part of the tangled natural world, with nothing to set them apart and no spark to fire their protection.

It was a humbling thought that neither Grant, Phil, Katherine or I would have been discussing the fate of a beetle or an ant at the William Creek pub. Although the insects of the Davenport Ranges are more important to the regional ecology than the wallabies, none has a 'story'. In management and public opinion terms they are irrelevant.

I remembered a bizarre-looking ant with black spikes all over its back and a glowing golden–orange bum. These distinctive ants crawl all over the Davenports and I had also seen them on other ranges to the north. However they did not appear in my field guide to South Australian ants. Alan Andersen, the mentor for many Australian ant researchers, later confirmed my suspicion that they were a type of *Polyrhachis*. They probably didn't have a name and they definitely did not have a story, or not one that was accessible to me anyway. This was arguably the most distinctive ant in the Davenports yet no-one knew anything about them. There are possibly hundreds or thousands of unglamorous species or habitats in the outback suffering similar problems to the wallabies but no-one knows enough about them or their plight to do anything about it.

A rare and noteworthy exception is the series of Lake Eyre mound springs which are now rightly regarded as unique and important environments. One hundred years ago the mound springs were important to explorers, the last of the nomadic Aboriginal people and the early pastoralists who all relied on the springs for water. The Aboriginal people burnt the dense reedbeds to maintain water flows and the springs were fenced to prevent stock from destroying them. But following the drilling of bores that provided higher flow rates in better grazing country, the importance of these springs was diminished and the fences fell into disrepair. Cattle and sheep pugged and churned them up into quagmires like the one I had experienced at Fred Spring with the bogged cow. No-one cared about the redundant mound springs. They neither had a use nor anyone with the motivation to protect them and as a result many 'died'.

Fortunately the status of the springs improved again in the late 1970s. A survey by the South Australian Nature Conservation Society, which was prompted by the possibility of future water extraction for the proposed Olympic Dam mine, alerted Wolfgang Zeidler, from the South Australian Museum, to a range of interesting invertebrates in the springs. Wolfgang, along with Winston Ponder, a snail boffin from the Australian Museum, quickly described many species of snails and small crustaceans that were new to science. Although several generations of pastoralists and travellers barely noticed the tiny poppy seed-like snails that Winston and Wolfgang found fascinating, all of a sudden their story was told and the mound springs resumed their status as important and interesting ecosystems.

Since the initial flurry of interest, over 120 springs have now been included within National Parks and others, on pastoral land, have been fenced off from grazing. WMC alone has spent around $10,000,000 managing and researching the Great Artesian Basin and its special ecosystems. These springs in South Australia are now better managed and their unique plants and animals more appropriately safeguarded than twenty years ago. Winston still laments that many of the mound springs in Queensland have not been surveyed for invertebrates although they are seriously affected by pastoral water extraction and grazing. They lack a high profile threat like the controversial mining venture which opened up the story on the South Australian springs.

The little invertebrates at the heart of Wolfgang and Winston's springs are not necessarily any more biologically significant than the orange-bummed ants or countless other little critters in the outback. One day we may discover that some of the spiders, water beetles or algae that live in the springs may have an even more compelling story than the snails. It is difficult to feel responsible and even harder to feel guilty for destroying something we don't know about. Fortunately, if we set out to conserve icon species like wallabies, pretty birds or hydrobiid snails, then the orange-bummed ants and countless other organisms will be protected in the process.

Few people are motivated to take action on behalf of a species that doesn't have a story to tell. There are also not nearly enough storytellers, National Parks managers, corporate donors or hard-working volunteers to stabilise the applecarts throughout the country. None of us can sit back and say that 'they' are responsible for the erosion of our natural heritage. 'We' are all as much to blame and, as the rock wallabies showed, if we are not motivated into action we will continue to lose species and Australia will become a poorer place.

28
Looking forward to the past

Mid-February is synonymous with Valentine's Day, the time of the year for quiet romantic evenings. But February 13 in 1999 was the date of one of the wildest, most exuberant parties that I have ever experienced. Together with Katherine and a close-knit group of friends, I was running amok within sight of the newly erected Olympic Dam smelter stack. However, we were not celebrating the completion of this massive industrial project, we had a far more scompelling reason to kick up our heels in the red sand. We were celebrating a hard-won environmental victory.

The bones of twenty rabbits that had been satayed, stewed and curried littered the ground around our camp ovens. Never before had 'underground mutton' tasted so good. A glitter ball suspended from a mulga branch erratically reflected the beams of our head torches as we danced around like dust-coated primeval warriors on the balmy still night. Status Quo and The Living End blared from a portable CD player and elevated our spirits even higher. Melted ice and spilt beer quickly turned the makeshift dance-floor into a mud pit, which provided further outlets for our peaking satisfaction and excitement.

We were experiencing the joy of eventual success after suffering so many trials and obstacles, false starts and frustrations. For a couple of years we had been reaching for a goal that we had been told was unachievable. Even those of us in the thick of the action had resigned ourselves to the fact that our target was unrealistic and the odds were too great.

Since RCD had worked its wonders Katherine and I had dreamt of restoring a part of the outback to the state it enjoyed before being ravished by introduced animals. For the first time since myxomatosis, low rabbit numbers had provided an opportunity to control feral animals from a large enough area to allow the reintroduction of locally extinct animals.

Keith Greenfield had told us that his father remembered seeing bilbies near where I had found old stick-nest rat nests and where Max Hall had seen woma pythons, just to the south of Roxby. The biggest clue to the types of native animals that lived in the district was stumbled onto by Darren Niejalke, who had helped me with surveys of Lake Eyre and the desert frogs, and his partner Helen Owens, who had found the guinea pig near Dalhousie. Hels and Darren found a deposit of semi-fossilised bones under some boulders only forty kilometres from Roxby. This deposit indicated the past presence of a staggering sixteen species of mammals that were no longer present in this region. Maybe it was possible to bring those animals that were still found in isolated refuges back to their rightful home. We threw around ideas about the possibility of creating a huge sanctuary near Roxby Downs.

Those long walks out to the Lake Eyre islands reminded me of the biogeographical principles that a 'sanctuary' had to be large in order to maintain a diverse range of species. The patchiness of local pine and western myall germination also showed that a reserve in this type of country had to be large enough to intercept patchy, yet incredibly important rains. From her work with rare rodents Katherine knew the importance of protecting areas that could support several populations of animals.

Before we applied for permission and funding from WMC, who had previously indicated that they were interested in such a project, Katherine and I decided to secure some key support. Pete Copley from the South Australian Department of Environment was an obvious point of contact. Katherine had worked with Pete on the reintroduction of several species of rare mammals on off-shore islands and I knew him from several field trips and the research into desert mice and plains rats. Pete was characteristically cautious but he indicated that a long-term research-based conservation project, with the Department of Environment as an integral participant, would receive his support. Our ecological mentor, Dave Paton, agreed that Adelaide University could provide the research base for a large, jointly managed project. As Katherine had worked for Pete, Dave and WMC over the previous few years, she was in a good position to prepare the proposal.

Katherine scoured the local countryside by foot and air for a few days, checking out potential sites with good access, security and that allowed for the possibility of expanding the sanctuary in the future. Eventually she settled on a site about five kilometres north of the mine in an area that had only ever been lightly grazed by cattle. The area incorporated dunes, gibber flats, woodlands and swamps, and fitted all of the other criteria that we had agreed on.

We budgeted on raising $150,000 from WMC over a two-year construction phase. If we tapped into additional funds and plenty of volunteer labour, we figured that we could afford fourteen kilometres of fencing. This would get us started while rabbit numbers were low, although we recognised that in the long-term the sanctuary should ideally be much larger.

To protect as much area as possible the newly named Arid Recovery Reserve should be square in shape. Limiting the number of corners or kinks also minimised expenses and weak points in the fence. However there were endless reasons why we could not build four straight sides. Wherever possible, fences should cross dunes at their lowest points and avoid swamps and tree groves. A straight fenceline that avoided high dunes, swamps and trees at one end would invariably be inappropriate at the other end. Therefore, starting with a fourteen-kilometre line on an aerial photo, followed by weeks of walking half marathons along potential fence alignments, Katherine narrowed down the best location for the fence.

Willy Williams along with his mates Jimmy Bannington and Willy Dingle, who had cracked the gengelope mystery for me, were keen that we bring back animals that they had seen as boys or knew through stories or songs. We sought their approval before clearing the fenceline. Perhaps more than any other environmental or heritage issue, the existence and value of sacred or significant Aboriginal sites arouses suspicion in whitefellas intending to build fences, bridges, mines or anything else. However I had a lot of respect for these old blokes after an amazing experience that I shared with them a few years earlier.

I had been with them as they slowly followed an unmarked 'songline' near Roxby in an area that they probably had never been before and had certainly not visited for at least twenty years. Observing them negotiating an apparently unmarked trail known to them only by stories handed down by their ancestors is truly humbling. With hands clasped behind their backs and bowed heads oblivious to the swarming flies, they quietly hummed ancient songs that guided them through this unfamiliar landscape. After half an hour or more, as if drawn by an invisible magnet, the men made a beeline along a cleared trail on a gravelly claypan towards a couple of head-sized rocks. I was summoned over and shown a rock that was crudely, but unmistakably engraved with a face. I never found out the significance of that face in their culture. However, its importance for me was that these magnificent old *tjilpis* had been aware of its presence and were able to locate it using a timeless, musical roadmap and an innate bonding with the land. No-one who had the opportunity to observe the old men on that day would have any doubt about the authenticity of their claims.

Many of the optimal low sites for the fence to cross dunes were blowouts that were also favoured localities for the ancestral Aboriginal people to shape rocks into cutlery. Willy Williams shook his head when we asked if one particular scattering of rocks was important. 'Movim outatha way, palya', he mumbled as he flicked a couple of artefacts away from the proposed fenceline. Other areas, including some sites where we had not noticed any unusual rocks, were important to them and we adjusted the fenceline accordingly, adding in extra kinks each time.

The thousands of rock artefacts indicated that the proposed reserve was once a popular location for the nomadic Aboriginal people in good seasons. Despite their love of kangaroo meat it is likely that these people traditionally ate mainly plant material. The artefacts on the dunes included large flat stones and fist-sized round grinding stones, worn and pitted through years of use. I asked Willy what type of seeds were ground with these rocks, expecting him to point out the wheat-like Mitchell grass or acacia seeds. It was not until a few days later that he could point out the type of grass seeds that were used. Amazingly, despite their tiny seeds and current sparse distribution, pannicum seeds were apparently a major food source for the nomadic Aboriginal people. Possibly as a result of rabbits or other recent changes to the regional ecology, this grass was now far less common than it must have been when grind stones were the vogue kitchen appliance.

Once we worked out where we could build the fence we had to settle on the optimal design to keep out rabbits, cats and foxes. The first Australian rabbit-netting fences were built in the 1840s and by 1914 a staggering 320,000 kilometres of fencing had been erected in a futile attempt to control rabbits. Therefore as far as the rabbit component was concerned we did not have to reinvent the wheel, or so we thought!

The unlikely Australian champion of the more elaborate and expensive cat- and fox-proof fences is an ex-mathematics lecturer from Flinders University. John Wamsley is South Australia's most recognised contemporary 'environmentalist'. His reputation has been earned as much for his controversial methods for attracting media attention and for denouncing conventional endangered species 'management', as for his achievements with breeding endangered mammals. Following the success of his effective yet expensive prototype cat-proof fence at Warrawong Sanctuary, John has modified his design for subsequent fences at his other Earth Sanctuaries.

Because we were aiming to construct the longest fence possible with our sponsorship money, we were keen to build the cheapest and most effective rabbit-, cat- and fox-proof fence. After Katherine had looked at a range of designs including those of Earth Sanctuaries, she settled on the fence design used recently on Wardang Island to separate the RCD-trial rabbits from the abundant feral cats. This two-metre high fence had a non-rigid overhang, or 'floppy-top' that apparently prevented cats from climbing over. The bloke who built the Wardang Island fence, Ray Wallace, claimed that it was absolutely rabbit and cat proof and, furthermore, he was available to supervise the construction of our fence if we could arrange the finances. A fence that was cat and rabbit proof but which leaked RCD seemed like our perfect solution. We had attracted key partners, a prime location, heritage approval, a fence design and had put together a budget. We also had a prominent yet unlikely ally within the hierarchy of WMC, whom five years earlier had made a statement that we hoped would help us to secure funding for the project.

Hugh Morgan AO, or 'Huge' as he is colloquially referred to when out of earshot, is the Chief Executive Officer of WMC. Despite his tongue-in-cheek nickname, Huge is no burly mountain of a man. Instead, his choirboy fringe and cavernous smile are carried on a small frame. Despite Huge's somewhat meagre physical stature he is a giant figure in corporate Australia with an imposing intellect and a thorough devotion to the mining industry. By denouncing what he considers to be unnecessary impediments to industrial progress, Mr Morgan—as his Melbourne-based reports call him—has often been the nemesis of the environmental movement. Therefore, you can imagine my surprise when he suggested that the reintroduction of locally extinct conservation icons would support his assertion that the mining industry had come of age, and that it was now acting in an environmentally appropriate manner.

Unfortunately our timing to put Huge's idea to the test could not have been worse. Low gold prices had forced the closure of several WMC mines in Western Australia. Low copper prices also meant that the Olympic Dam mine was not the shining knight that it was hoped to be. Our belts were tightened and the staff 'downsized'. Long-term employees, some of whom had been my mates for many years, were given their marching orders. Senior managers were cautious about backing the project without the unanimous support of other key managers. This was hardly the time to be asking for funds for an environmental flight-of-fancy. But RCD was not affected by metal prices and its effects had given us a chance to achieve something special.

After a great deal of lobbying the company tentatively authorised us to proceed with the fencing of the initial fourteen square kilometre reserve. WMC would provide the seed money to start the fencing while the rabbit numbers were low. However the Olympic Dam manager, Dave Thomas, was particularly concerned that we were planning to build an expensive paddock for a giant white elephant. His authority to proceed was couched with the instruction that WMC would provide 'no additional funding, no airfares, no lunches, nothing!' Furthermore, we were told that if it was the great project that we had sold them on, we should be able to raise half of the necessary funds from elsewhere. Given that the University and National Parks are perpetually cash-strapped we initially thought that this caveat would suffocate the project, but in hindsight it was just the challenge that we needed. Our trump card was that we could include volunteer labour in the equation.

The Sedan footy club, under the persistent supervision of their club stalwart Ray Wallace, agreed to erect the fence on their end-of-season trip at a fraction of the cost of professional fencers. They were joined by a procession of locals who were captivated by the project and offered assistance after work and on weekends. Fortunately time softens the memory of the arduous and frustrating task of erecting a six-foot high netting fence, and my recollections of the two-month construction phase were of premierships to my local basketball side, my beloved Redlegs footy team and the phenomenal first AFL flag for the Adelaide Crows. Despite the euphoria of these wins nothing can fully erase the memory of all of those bloody fence clips.

Building conventional fences is generally a rewarding task. At the end of the day you can clearly see the progress made as a line of posts and droppers steadily marches out across the countryside. Clipping netting on to the fence is another matter. The boys from Sedan along with the Roxby volunteers had to secure the horizontal foot netting to the bottom wire every five centimetres with staple-like clips. This would then prevent the rabbits from burrowing under the fence. The temperamental air-clip guns frequently seized up, particularly when they were cluttered with the debris from clipping the bottom wire. Paragraphs of verbal abuse were cursed at the manufacturers of the clip guns, who repeatedly returned them to us with a note saying that they were working well despite the fact that they would jam again within a few minutes. Several times during the construction of the fence we seriously wondered whether we would ever make it. Even after the fence was erected we still had the laborious task of adding two electric wires and inserting support wires into the 'floppy' overhang on the top of the fence.

Even before the last of the 500,000 clips were cussed into the netting we started removing the rabbits. As if in retribution for the role that feral animals had in the demise of the wallabies, Katherine attacked the rabbit eradication with fanatical gusto. With a contingent of volunteers she scoured the entire Reserve several times, fumigating and collapsing all warrens with spades. Volunteers zigzagged along the dunes where most of the rabbit holes were found. Many of the helpers could only work occasionally while others were not able to cope with the seemingly overwhelming task, day after day for months on end. Fortunately there were two notable exceptions.

Jackie Bice had recently graduated from a natural resource management course after working at the Adelaide Zoo for several years as an animal keeper. She stood out in the mining town with her greenie stickers and nose stud—a year or so before facial piercing was more common up here. Like most graduates, Jackster was aware that a key to obtaining work was to first demonstrate her value as a volunteer, and her beaming smile and unrelenting work ethic quickly earned her a couple of year's employment with the project. Jackster's love of the bush and enthusiasm for the exciting new project was exactly the support that her mate Katherine desperately needed.

The next volunteer to demonstrate his value and earn a job on the rabbit eradication team was Freebie. To picture Freebie, think back to the 'bad boy' of Australian long jumping at the Sydney Olympics, Jai Taurima. With lanky dark hair and even longer lankier limbs, Freebie, like Jai, was more at home on a surfboard than doing anything resembling study, training or work. But along with a passion for the surf around his Port Lincoln home, Freebie derived a form of primeval satisfaction in 'controlling' feral animals. He strived tirelessly to track down any rabbit that dared to live on the dunes assigned to him and then erupted with unbridled emotion when he had killed it. For the girls, killing rabbits was an unpleasant but necessary means to an end, but Freebie thrived on it. As a result, he quickly earned his 'natural born killer' title. With Katherine, Jackster and Freebie eradicating rabbits full-time on their designated dunes, we thought that it was only a matter of time that we could declare our newly fenced reserve to be rabbit free. How wrong we were.

※ ※ ※

As the sand gradually warmed from a grey–red to a glowing orange with the rising sun, three or four sets of boots would methodically zigzag over the dunes, contributing to the 'canvas' of messages left by insects, mice, lizards, snakes and birds. With up to ten kilometres of dune to cover before the elevated sun obscured the faint tracks, it generally took a week or more to methodically cover the allotted dunes and inspect large bushes or mounds that might harbour a rabbit hole. We then fumigated, blew-up, caved-in or blocked off occupied warrens. We also shot rabbits by day and night, poisoned them with 1080-laced oats and set soft-jaw and cage traps for particularly canny individuals.

At first the rabbit onslaught was depressingly unrewarding. Despite accounting for over 1,000 rabbits their tracks were still everywhere. Some rabbits didn't eat oats and others didn't live in warrens or use buck heaps where they could be trapped. But persistence finally paid off and the rabbits were confined to discrete pockets. Tactics then changed and several of us would target individual rabbits by following their tracks until we flushed them or found their fresh hole.

After tracking the same rabbit on several consecutive mornings we often found that its nocturnal movements followed a predictable pattern. 'A rabbit with a habit is a dead rabbit', was the catchcry as soon as a rabbit started to regularly use a buckheap or a runway between favoured feeding grounds. Once their habit was cracked the rabbit was inevitably accounted for within days. However, a few rabbits were far more difficult.

One particularly annoying rabbit left its tracks near the only gate into the Reserve, right where we had to look first thing each morning and last thing in the afternoon. Finding rabbit tracks was a real insult, like seeing rubbish on the ground after 'Clean-up Australia Day' or a bird turd on the windscreen just after you have cleaned your car. This rabbit, known as Gateboy, assumed a complex personality even though no-one had actually seen him, or her. So many times Katherine, Jackster or Freebie would confidently declare that they had trapped Gateboy or fumigated his hole only to find his tracks again the next day. He was a sly, calculating, sinister rabbit.

Brian Cooke, the rabbit researcher who brought the fleas to Roxby and RCD to Australia, told us that rabbits generally move only a few hundred metres from their warren. We found that they were much more mobile after becoming battle-hardened, poison-resistant guerrillas. Brian had obviously not included the likes of Gateboy in his studies. Once their warrens were destroyed and their neighbours removed, Gateboy and the last few rabbits were tracked for several kilometres every day.

Another rabbit that we learned to despise and admire in a strange way was Bunston. Following the eventual demise of Gateboy and the other rabbits our entire efforts eventually focussed upon this one animal. Bunston was the bunny equivalent of James Bond. No matter what we tried we couldn't get him. He refused to develop a habit of running in the same place on successive days, did not live in holes or use buck heaps, could not be tempted by oats, and his tracks had a knack of disappearing onto hard clay whenever it seemed like we were closing in on him. No matter how we tried to outsmart him Bunston was our sole nemesis for over a month.

The date of Bunston's demise was one of the most unusual days I ever witnessed at Roxby. Until nearly lunchtime the landscape was blanketed in an incredibly thick fog. My first thoughts were with Pete Paisley who organised Roxby's biggest annual party called Boogie in the Bush. I couldn't believe how unlucky Pete was to have landed a fog-bound day for the party, since we seldom even experience a light fog for more than a couple of hours every few years. However, within hours I couldn't believe how lucky we were.

The fog cleared and Freebie was able to check whether Bunston was still at large. Finally, perhaps baffled by the fog, we had captured the ingenious Houdini of the rabbit world. In fact 'he' turned out to be a 1335.8-gram female! Despite mid-July being in the peak breeding season for rabbits Bunston was not pregnant or lactating, which we saw as an indication that she did not have access to any males. Celebrating Bunston's demise at Boogie in the Bush was made even better by the realisation that our vigil of checking for rabbits first thing every morning had finally ended.

Exactly a month after the demise of Bunston, Bill Clinton announced to the world and his wife that he had been involved in an inappropriate relationship with a girl named Monica. On the same day Freebie, who later had a more appropriate relationship with a more appealing Monica, made a disturbing announcement. He had just finished a final check of all of 'his' dunes within the Reserve and had found a young rabbit. How could it be? Bunston, the barren female, was our indication that there were no more rabbits inside, especially not a breeding pair. Katherine had booked her tickets for a well-earned Alaskan holiday content that the rabbits had gone. A week after Freebie had tracked down and removed his interloper, Jackster found two more rabbits on dunes that had been 'clean' for months. Her ballistic screams of anguish echoed across the land. And then there were more.

A year had passed since we started the assault and we still had rabbits. What was arguably the largest, most thorough eradication program for rabbits in Australia's history had not worked. Maybe it couldn't. Katherine and Jackster were exasperated. Tantrums and tears joined the sweat that pattered small divots in the orange dust. Even the indomitable Freebie was shaking his curtain of surfy hair. There were no holes under trees on the dunes where the rabbits typically burrow. Each bush and tree had been checked countless times. The rabbits had to be coming from a warren that we had somehow overlooked on the wide swales between the dunes. We rechecked the swales, combing them as if we were homing in on Lasseter's Reef, rather than an elusive rabbit hole. Nothing. Then we made an alarming discovery.

One morning in October, Freebie and I were hot on the trail of a kitten rabbit that had been noticed a few days earlier. It definitely had not developed a habit and we had followed the little fella for a couple of kilometres before we flushed it from under a bush. It was tiny, only about the size of a rat and it had as much chance of outrunning us as we had of running down its mum. It doodled around between us, bumbling around vainly until we both dived on it.

It was hard for us to believe that the rabbit tracks we had followed up and down the dunes belonged to the limp little body in our hands. Surely a rabbit like this would not yet be weaned. But it must be, we had not seen any other tracks with it. Hanging by its ears and without the musculature of life, it looked so scrawny. Then the realisation dawned. This little rabbit might be able to get through our fence. Maybe we had built a fence that was not rabbit proof. Maybe Bunston was the last of the originals and we had been chasing transgressors ever since? We raced over to the fence and performed the unthinkable. The limp form, weighing in at only 168 grams, slipped straight through the mesh. The next rabbit we caught a few days later was twice its weight and was definitely independent of its mother, as we had been chasing its lone tracks for days. Once again it pulled straight through the fence like a pipe cleaner.

Katherine and Jackster were rightfully sceptical. A limp dead rabbit being pulled through a fence by its ears was one thing but could a live rabbit breach the fence? A couple of days later they had an opportunity to test this theory. Another live young rabbit that they tracked down and caught was put near the fence. After an initial nose twitch it casually hopped straight through the fence, turned around and hopped back again. They videoed the rabbit jumping through again and again. When the volume was turned up on the video, Katherine's expletives said it all. We had been duped. We had spent in excess of $200,000 and invested some of the most productive years of our lives on a fence that was not rabbit proof.

Ron Estens, the Southern Rural Manager for BHP wire products, viewed the video of the young rabbit hopping through the mesh and admitted that rabbits between 240 and 450 grams could get through the mesh that they marketed as 'rabbit proof'. However, because their netting performed according to their expectations and in line with the Parliamentary Acts concerning vermin control, they accepted no liability and still considered the netting to be 'rabbit proof'. Ron stated, 'We have been making this mesh for years and no-one else has ever complained that it is not rabbit proof.'

Waterproof watches stop water. Rabbit-proof netting should bloody well stop rabbits. Not just big rabbits or most rabbits, but all rabbits that are big enough to be independent. Big enough to eat plants, dig holes, grow up and make more bloody rabbits. If BHP had advertised their forty-millimetre mesh as 'forty-millimetre mesh that is useful at keeping out large rabbits', we would not have had a gripe. We could even understand if somewhere in fine print they mentioned, 'Independent rabbits up to 450 grams can penetrate this mesh'. Every one of the 300 fifty-metre rolls of useless mesh we had clipped onto our fancy but ineffective fence was wrapped in bright orange paper that exclaimed in big letters 'Rabbit-proof Mesh'. No ifs or buts about it.

To their credit BHP came to Roxby to face the fire. They sympathised with our problem, offered to provide us with smaller thirty-millimetre mesh and gave us a donation to offset its erection costs. They wouldn't come at reimbursing us for the extra months of rabbit control, not to mention the grief that we endured because their mesh leaked. In the bizarre manner of corporate politics we were told that their donation was a goodwill gesture, rather than an admission of phoney marketing.

To rub salt into the wound, a couple of months later when asked whether they were going to change their promotion for the mesh, or start advertising the smaller thirty-millimetre mesh as rabbit proof, they said no. Three years later you could still read the big orange and black lies on rolls of forty-millimetre mesh. Not only had we been duped but so too had countless other Landcare, Greening Australia, Soil Conservation Board, National Park and private conservation projects throughout the country.

By the time we had refenced with the thirty-millimetre mesh the rabbits had been exterminated. We made doubly sure that they could not reinvade by laying used conveyor-belt rubber next to the fence.

This would prevent moving dune sand from exposing the foot-netting that stopped rabbits from burrowing under the fence. A week later we celebrated our achievement with the now legendary rabbit party.

A year or so later, Katherine and I walked along the crest of 'dune 20' within the Arid Recovery Reserve. A wedgey nest spilled from the gnarled branches of an ancient weather-beaten pine. Old bleached rabbit skulls littered the sand underneath but there were no other traces of rabbits in the whole Reserve.

The local community were still supportive of the project and many thousands of dollars of donations and volunteer labour had been vital to its success. WMC had also been absolutely convinced of the Reserve's worth and had loosened their purse-strings to enable us to fast-track our planned developments. Neighbouring pastoralists donated parts of their leases to be included in a larger Reserve. National and international conservation bodies were now backing the project that was attracting research students from several universities. Katherine and I knew that the unheralded support from so many different people and organisations had cleared the slate for the first time in over a century. For a rare, brief moment we were both content.

We thought forward to the tracks of native species that might in time replace the rabbit tracks that we had grown to despise. Katherine wondered if the dream of bringing bilbies and other endangered critters back to their country could now become a reality. I thought back to what old Willy Williams had told me about the panicum grass, and wondered what the vegetation of the Reserve would look like in twenty or fifty years time. Already wild plum saplings and native apricots were growing prolifically, freed from cattle and rabbit grazing. Seedling mulgas and masses of crimson Sturt's Desert peas had already smothered the champagne cork where I proposed to Katherine a few months earlier.

We continued walking to the fence. On the outside were the tracks of a rabbit that had hopped up to a rattlepod plant and dug out its roots. Reality snapped us from our self-congratulatory trances. What we had achieved was just the beginning. We had shown that the number one enemy of the outback could be eliminated from a fortified paddock through outrageous expense and dedication. Now, largely due to the efforts of an ingenious bushy named Greg 'Mr Fixit' Kammerman and Greencorp volunteers led by the super-energetic pocket rocket, Nicki Munro, a further thirty-three kilometres of fencing were under construction. Even with this quadruple-sized reserve we knew that long lines of mesh linked by an infuriating number of wire clips could not achieve our aims on a broad scale.

Hopefully the continued commitment of mining, pastoral, government, research and tourist organisations will help to bring back a thriving ecosystem to the outback. However the real key to atoning for the losses of *Caloprymnus*, the rock wallabies and the many other species that are still struggling lies with action by informed and motivated local and national communities. No amount of talking, planning and strategies can replace sweat, blisters and persistent lobbying for change. We can't afford to put our feet up, or take our hats off yet.

Bibliography

Texts pertaining to more than one chapter are only listed once, in the chapter of most relevance. More information on the topics covered in this book, links to relevant websites and the author's contact can be found at www.redsandgreenheart.org

Chapter 1

Curtis, H (1983), *Biology* 4th Edition, Worth Publishers, New York.
Elton, CS (2000), *The ecology of invasions by animals and plants*, University of Chicago Press, Chicago.
Gilmour, AR, McDougall, KW & Spurgin, P (1999), 'The uptake and depletion of fenitrothion in cattle, pasture and soil following spraying of pastures for locust control', *Australian Journal of Experimental Agriculture*, no. 39, pp. 915–22.
Story, P & Cox, M (2001), 'Review of the effects of organophosphorus and carbamate insecticides on vertebrates. Are there implications for locust management in Australia?', *Wildlife Research*, no. 28, pp. 179–193.
York Main, B (1982), 'Adaptations to arid habitats by mygalomorph spiders', *Evolution of the flora and fauna of arid Australia*, Barker, WR & Greenslade, PJM (eds), pp. 273–283, Peacock Publications, Adelaide.

Chapter 2

Carr, SG & Robinson, AC (1997), 'The present status and distribution of the Desert Rat-kangaroo *Caloprymnus campestris* (Marsupialia: Potoroidae)', *South Australian Naturalist*, no. 72, pp. 4–27.
Finlayson, HH (1961), 'On Central Australian Mammals. Part IV-The distribution and status of central Australian species', *Records of the South Australian Museum*, no. 14, pp. 141–191.
Miller, AH (1963), 'The fossil flamingos of Australia', *The Condor*, no. 65, pp. 289–299.
Morton, SR (1990), 'The impact of European settlement on the vertebrate animals of arid Australia: a conceptual model' *Proceedings of the Ecological Society of Australia*, no. 16, pp. 201–213.
Olson, SL & Feduccia. A (1980), 'Relationships and evolution of flamingos (Aves: Phoenicopteridae)', *Smithsonian Contributions to Zoology*, no. 316.
Short J and Smith (1994), 'A mammal decline and recovery in Australia' Journal of Mammalogy, no. 75, pp. 288–297.

Chapter 3

Read, JL (1997), 'Stranded on desert islands? Factors shaping animal populations in Lake Eyre South' *Global Ecology and Biogeography Letters*, no. 6, pp 431–438.

Chapter 4

Allan, RJ, Bye, JAT & Huton, P (1986), 'The 1984 filling of Lake Eyre South' *Transactions of the Royal Society of South Australia*, no. 110, pp. 81–87.
Dalhunty, JA (1974), 'Salt crust distribution and lake bed conditions in southern areas of

Lake Eyre North' *Transactions of the Royal Society of South Australia*, no. 98.
Dalhunty, R (1975), *The spell of Lake Eyre*. Kilmore, Lowden Publishing Co.
MacArthur, RH & Wilson, EO (1967), *The Theory of Island Biogeography*, Princeton University Press, Princeton N.J.
Read, JL (1996), 'Fauna of the Elliot Price Conservation Park.' *South Australian Naturalist*, no.69, pp. 4–11.

Chapter 5
Black, JM (1988) 'Preflight signalling in swans: a mechanism for group cohesion and flock formation', *Ethology*, no. 79, pp. 143–57.
Morton, SR, Doherty, MD & Barker, RD (1995), *Natural heritage values of the Lake Eyre basin in South Australia: World Heritage Assessment*, CSIRO Wildlife and Ecology, Canberra.
Read, JL & Badman, FJ (1999), 'Birds of the Lake Eyre South region', *Lake Eyre South Monograph Series*, Slater WJH (ed.), Vol. 3, RGSSA, Adelaide.
Ruello, NV (1976), 'Observations on some massive fish kills in Lake Eyre', *Australian Journal of Marine and Freshwater Research*, no. 27, pp. 667–72.
Vestjens, WJM (1977), 'Breeding behaviour and ecology of the Australian Pelican, *Pelecanus conspicillatus*', in *New South Wales Australian Wildlife Research*, no. 4, pp. 37–58.
Waterman, MH & Read, JL (1992), 'Breeding success of the Australian Pelican *Pelecanus conspicillatus* on Lake Eyre South in 1990', *Corella*, no. 16, pp. 123–126.

Chapter 6
Haig, SM, et al, (1997), 'Population identification in shorebirds', *Molecular Ecology*, no. 6, pp. 413–427.
Marzluff, JM, Heinrich, B & Marzluff, CS (1996), 'Roosts are mobile information centres', *Animal Behaviour*, no. 42, pp. 89–103.
Walraff, HG & Hund, K (1982), 'Homing experiments with starlings (*Sturnus vulgaris*) subjected to olfactory nerve section', *Avian Navigation 1982*, pp. 313-318.

Chapter 7
Dorfman, E and Read, JL (1996), 'Observations of nest predation by corvids on cormorants in Australia and the potential for population regulation' *Emu*, no. 96, pp. 132–35.
Marchant, S & Higgins, PJ (1993), *Handbook of Australian, New Zealand and Antarctic Birds*, Volume 2: Raptors to Lapwings, Oxford University Press.
Read, JL & Ebdon, R (1998), 'Waterfowl of the Arcoona Lakes: An important arid zone wetland complex in South Australia', *Australian Bird Watcher*, no. 17, pp. 234–244.

Chapter 8
Read, JL (1999), 'A strategy for minimising waterfowl deaths on toxic ponds', *Journal of Applied Ecology*, no. 36, pp. 345–350.
Weidensaul, S (1999), *Living on the wind: Across the hemisphere with migratory birds*, North Point Press, New York.

Chapter 9
Read JL (1995), 'Recruitment characteristics of the White Cypress Pine (*Callitris glaucophylla*) in arid South Australia', *Rangeland Journal*, no. 17, pp. 228–240.
Showers, J (1999), *Return to Roxby Downs*, Bookend Books, Unley.
Stafford-Smith, DM & Morton, SR (1990), 'A framework for the ecology of arid Australia', *Journal Arid Environments*, no. 18, pp. 255–78.

Chapter 10
James, CD (1991), 'Population dynamics, demography, and life history of sympatric scincid lizards (*Ctenotus*) in central Australia' *Herpetologica*, no. 47, pp. 194–210.
James, CD, Landsberg, J & Morton, SR (1995), 'Ecological functioning in arid Australia and research to assist conservation of biodiversity', *Pacific Conservation Biology*, no. 2, pp. 126–42.
Kotwicki, V (1986) *Floods of Lake Eyre*, Engineering and Water Supply, Adelaide.
Read, JL (1992), 'Influence of habitats, climate, grazing and mining on the small terrestrial vertebrates at Olympic Dam, South Australia', *Rangeland Journal*, no. 14, pp. 143–156.
Read, JL (1999), 'Longevity, reproductive effort and movements of three sympatric Australian arid zone gecko species', *Australian Journal of Zoology*, no. 47, pp. 307–316.
Read, JL & Badman, FJ (1990), 'Reptile densities in chenopod shrubland at Olympic Dam, South Australia', *Herpetofauna*, no. 20, pp. 3–7.

Chapter 12
Broad, AJ, Sutherland, SK & Coulter, AR (1979), 'The lethality in mice of dangerous Australian and other snake venoms', *Toxicon*, no. 17, pp. 661–664.
Covacevich, J (1987), 'Two taipans!', *Toxic plants and animals: A guide for Australia*, Covacevich, J Davie, P & Pearn, J, Queensland Museum, Brisbane, pp. 481–485.
Ludwig, JA (1987) 'Primary productivity in arid lands: myths and realities', *Journal of Arid Environments*, no. 13, pp. 1–7.
Read, JL (1992), 'Ecological and biological notes on the rare plant *Hemichroa mesembryanthema* F. Muell (Amaranthaceae)', *Transaction of the Royal Society of South Australia*, no. 116, pp. 145–146.
Read, JL, Bird, B & Greenfield, C (1996), 'Southern range extensions of Flock Bronzewings *Phaps histrionica* in South Australia including a breeding record', *South Australian Ornithologist*, no. 32, pp. 99–102.
Reese, LR (1924), 'Bird Notes', *South Australian Ornithologist*, no. 7, pp. 229–230.
Reid, J and Fleming, M (1992), 'The conservation status of birds in arid Australia', *Rangeland Journal*, no. 14, pp. 65–91.
White, J (1987), 'Elapid snakes: Venom toxicity and actions', *Toxic plants and animals: A guide for Australia*, Covacevich, J, Davie, P & Pearn, J, Queensland Museum, Brisbane, pp. 369-389.

Chapter 13
Brandle, R, Moseby KE & Adams M (1999), 'The distribution, habitat requirements and conservation status of the plains rat, *Pseudomys australis* (Rodentia: Muridae)', *Wildlife Research*, no. 26, pp. 463–477.
Mirtschin, PJ & Reid, RB (1982), *Transactions of the Royal Society of South Australia*, no. 106, pp. 213–214.

Read, JL (1994), 'A major range extension and new ecological data on *Oxyuranus microlepidotus* (Reptilia: Elapidae)', *Transaction of the Royal Society of South Australia*, no. 118, pp. 143–145.

Read, JL and Owens, HM (1999), *Reptiles and amphibians of the Lake Eyre South region*, WJH Slater ed., Lake Eyre South Monograph Series, RGSSA, Adelaide, vol. 1, part 3.

Chapter 14

Bull, CM (1991), 'Ecology of parapatric distributions', *Annual Review of Ecological Systematics*, no. 22, pp. 19–36.

Cogger, HG, Cameron, EE, Sadlier, RA & Eggler, P (1993), *The action plan for Australian Reptiles*, Australian Nature Conservation Agency, Canberra.

Darwin, C (1859), *The origin of species by natural selection*, Avenel Books, New York.

Harvey, C (1981), 'A new species of *Nephrurus* (Reptilia:Gekkonidae) from South Australia', *Transactions of the Royal Society of South Australia*, no. 107, pp. 231–235.

Johnston, GR (1992), '*Ctenophorus tjantjalka*, a new dragon lizard (Lacertilia: Agamidae) from northern South Australia', *Records of the South Australian Museum*, no. 26, pp. 51–59.

Read, JL (1994), 'A retrospective view of the quality of the fauna component of the Olympic Dam Project Environmental Impact Statement', *Journal of Environmental Management*, no. 41, pp. 167–185.

Read, JL (1995), 'Gull-billed Tern predation on dragons; a possible range limiting factor for Arcoona Dragons (*Ctenophorus fionni*)?', *Herpetofauna*, no. 25, pp. 50–53.

Read, JL (1998), 'Vertebrate fauna of the Nifty region, Great Sandy Desert with comments on the impacts of mining and rehabilitation', *Western Australian Naturalist*, no. 22, pp. 29–49.

Read, JL (1998), 'Hemmed in on all sides?: the status of the restricted gecko, *Nephrurus deleani*', *Herpetofauna*, no. 28, pp. 30–38.

Chapter 15

Mengden, GA (1995), 'A chromosomal and electrophoretic analysis of the genus *Pseudonaja*', *The biology of Australian frogs and reptiles*, Grigg, G, Shine, R & Ehmann, H (eds) Surrey Beaty & Sons, pp. 193–208.

Chapter 16

Ashton, CB (1996), 'Changes in the avifauna using Aldinga Scrub Conservation Park', *South Australian Ornithologist*, no. 32, pp. 93–98.

Blakers, M, Davies, SJJF & Reilly, PN (1984), *The Atlas of Australian Birds*, Royal Australasian Ornithologists Union, Melbourne University Press, Melbourne.

Higgins, PJ (1999), *Handbook of Australian, New Zealand and Antarctic Birds*, Volume 4: Parrots to Dollarbird, Oxford University Press.

Paton, DC, Carpenter, G & Sinclair, RG (1994), 'A second bird atlas of the Adelaide region. Part 1: Changes in the distribution of birds: 1974-75 vs 1984-85', *South Australian Ornithologist*, no. 31, pp. 151–193.

Pedler, L (1992), 'Review of the status and distribution of the Chestnut-breasted Whiteface *Aphelocephala pectoralis*', *South Australian Ornithologist*, no. 21, pp. 79–93.

Read, JL (1995), 'The ecology of the Grass Owl *Tyto capensis* south of Lake Eyre', *South*

Australian Ornithologist, no. 32, pp. 58–60.
Read, JL (1999), 'Bird colonisation of a remote arid settlement', *Australian Bird Watcher*, no. 18, pp. 59–67.
Read, JL, Ebdon FR & Donohoe, P (2000), 'The terrestrial birds of the Roxby Downs Area: a ten year history', *South Australian Ornithologist*, no. 33, pp. 71–83.

Chapter 17

Harfenist, A, Power, T, Clark, KL & Peakall, DB (1989), *A review and evaluation of the amphibian toxicological literature*, Technical Report Series No. 61, Canadian Wildlife Service, Headquarters.
Henle, K (1981), 'A unique case of malformations in a natural population of a green toad (*Bufo viridis*) and its meaning for environmental politics', *British Herpetology Society Bulletin*, no. 4, pp. 48–49.
Morton, SR, Masters, P & Hobbs, TJ (1993), 'Estimates of abundance of burrowing frogs in spinifex grasslands of the Tanami Desert, Northern Territory', *The Beagle*, no. 10, pp. 67–70.
Ouellet, M, Bonin, J, Rodrigue, J, Des Granges, J & Lair, S (1997), 'Hindlimb deformities (ectromelia, ectrodactyly) in free-living anurans from agricultural habitats', *Journal of Wildlife Diseases*, no. 33, pp. 95–104.
Predavec, M & Dickman, CR (1993), 'Ecology of desert frogs: a study from southwestern Queensland', *Herpetology in Australia, a diverse discipline*, Lunney, D & Ayers, D (eds), Transactions of the Royal Zoological Society of New South Wales, Sydney, pp. 159–169.
Read, JL (1997), 'Comparative abnormality rates of the Trilling Frog at Olympic Dam', *Herpetofauna*, no. 27, pp. 23–27
Read, JL (1998), 'The ecology of sympatric scincid lizards (*Ctenotus*) in arid South Australia', *Australian Journal of Zoology*, no. 46, pp. 617–629.
Read, JL (1999), 'Diet and causes of mortality of the Trilling Frog, *Neobatrachus centralis*', *Herpetofauna*, no. 29, pp. 2–7.
Read, JL (1999), 'Abundance and recruitment patterns of the Trilling Frog (*Neobatrachus centralis*) in the Australian arid zone', *Australian Journal of Zoology*, no. 47, pp. 393–404.
Read, JL (2000), 'Baseline abnormality study of the frogs of Tampakan, South Cotabato Province, Mindanao, Philippines', *Asia Life Sciences*, no. 9, pp. 1–6.
Read, JL & Tyler, MJ (1990), 'The nature and incidence of post-axial, skeletal abnormalities in the frog *Neobatrachus centralis* Parker at Olympic Dam, South Australia', *Transactions of the Royal Society of South Australia*, no. 144, pp. 213–217.
Read, JL & Tyler, MJ (1994), 'Natural levels of abnormalities in the Trilling Frog (*Neobatrachus centralis*) at the Olympic Dam Mine', *Bulletin of Environmental Contamination and Toxicology*, no. 53, pp. 25–31.
Tinsley, R & Tocque, K (1995), 'The population dynamics of a desert anuran, *Scaphiopus couchii*', *Australian Journal of Ecology*, no. 20, pp. 376–384.

Chapter 18

Moseby, KE & Read, JL (1998), 'Population dynamics and movement patterns of Bolam's mouse, *Pseudomys bolami*, at Roxby Downs', *Australian Mammalagy*, no. 20, pp. 353–368.
Read, JL (1996), 'Use of ants to monitor environmental impacts of salt spray from a mine

in arid South Australia', *Biodiversity and Conservation*, no. 5, pp. 1533–1543.
Read, JL (1998), 'Are geckos useful bioindicators of air pollution?', *Oecologia*, no. 114, pp. 180–197.
Read, JL, Copley, P & Bird, P (1999), 'The distribution, ecology and current status of *Pseudomys desertor* in South Australia', *Wildlife Research*, no. 26, pp. 453–462.
Read, JL, Kovac, K-J and Fatchen, TJ (forthcoming), '"Biohyets": a holistic method for demonstrating the extent and severity of environmental impacts', *Journal of Environmental Management*.
Read, JL, Reid, N & Venables, WN (2000), 'Which bird species are useful bioindicators of mining and grazing impacts in arid South Australia?', *Environmental Management*, no. 26, pp. 215–232.

Chapter 19

Australian Conservation Foundation (1999), 'Australia at the nuclear crossroads', *Habitat Australia*.
Cockburn, S & Ellyard, D (1981), *Oliphant. The life and times of Sir Mark Oliphant*, Axiom Books, Adelaide.
Hore-Lacy, I (1999), *Nuclear electricity*, Uranium Information Centre, Melbourne, Australia.
Read, JL & Pickering, R (1999), 'Ecological and toxicological effects of exposure to an acidic, radioactive tailings storage', *Environmental Monitoring and Assessment*, no. 54, pp. 69–85.
Sprigg, RC (1984), *Arkaroola-Mount Painter in the Northern Flinders ranges, S.A.: The last billion years*, Gillingham Printers, Adelaide.

Chapter 20

Amarasekare, P & Possingham, HP (2001), 'Patch dynamics and metapopulation theory: the case of successional species', *Journal of Theoretical Biology*, vol. 209.
Andrewartha, HG & Birch, LC (1954), *The distribution and abundance of animals*, The University of Chicago Press, Chicago.
Hughes, L (2000), 'Biological consequences of global warming: is the signal already apparent?' *Trends in Ecology and Evolution*, no. 15, pp. 56–61.
Quammen, D (1996), *The Song of the Dodo. Island Biogeography in an Age of Extinctions*, Pimlico, London.

Chapter 22

Baker, L, Woenne-Green, S & the Mutitjulu Community (1993), 'Anangu knowledge of vertebrates and the environment', *Uluru Fauna: The distribution and abundance of vertebrate fauna of Uluru National Park*, Reid, JRW, Kerle, JA & Morton, SR (eds), Australian National Parks and Wildlife Service, Canberra, pp. 79–132.
Bonython, E (1985), *Where the seasons come and go*, Illawong, Yankalilla.
Cockburn, R (1974), *Pastoral pioneers of South Australia*, Vol. 1, 1925 facsimile, Lutheran Publishing House, Adelaide.
Gee, P (2001), 'A history of pastoralism in the Lake Eyre South Drainage Basin', *Lake Eyre South Monograph Series*, Slater, WJH (ed.), RGSSA, Adelaide.
Landsberg, J, James, CD, Morton, SR, Hobbs, TJ, Stol, J, Drew, A & Tongway, H (1997), 'The effects of artificial sources of water on rangeland biodiversity', *Final report to the Biodiversity Group*, Environment Australia.

Phillips, A (1998), 'Cograzing cattle and camels for commercial production', *Annual report to RIRDC*, NT Department of Primary Industry and Fisheries.
Read, JL (1999), 'The initial response of a chenopod shrubland plant and invertebrate community to intensive grazing pulses', *Rangeland Journal*, no. 21, pp. 169–193.
Read, JL (2002), 'Experimental trial of Australian arid zone reptiles as early warning indicators of overgrazing by cattle', *Austral Ecology*, no. 27, pp. 55–66.
Read, JL & Andersen, AN (2000), 'Using ants as early warning indicators: Responses to pulsed cattle grazing in the Australian arid zone', *Journal of Arid Environments*, no. 45, pp. 231–51.

Chapter 23

Anderson, L (2000), *Genetic engineering, food, and our environment. A brief guide*, Scribe Publications, Melbourne.
Croft, DB (2000), 'Sustainable use of wildlife in western New South Wales: Possibilities and problems', *Rangeland Journal*, no. 22, pp. 88–104.
Denny, MJS (1982), 'Adaptations of the red kangaroo and euro (Macropodidae) to aridity', *Evolution of the flora and fauna of arid Australia*, Barker, WR & Greenslade, PJM (eds), Peacock Publications, Adelaide, pp. 179–183.
Grigg, G (1987), 'Kangaroos, a better economic base for our marginal grazing lands?', *Australian Zoologist*, no. 24, pp. 73–80.
Grigg, G (2002), 'Conservation benefit from harvesting kangaroos: status report at the start of a new millennium', *A zoological revolution: using native fauna to assist in its own survival*, Lunney, D & Dickman, C (eds), Royal Zoological Society of NSW, Mosman, pp. 53–76.
Hightower, R, Baden, C, Penzes, E, Lund, P & Dunsmuir, P (1991), 'Expression of antifreeze proteins in transgenic plants', *Plant Molecular Biology*, no. 17, pp. 1013–1021.
Lunney, D & Grigg, G (1988) 'Kangaroo harvesting and the conservation of arid and semi-arid lands', *Australian Zoologist*, no. 24, pp. 121–185.
Wilson, D & Read, JL (2003), 'Kangaroo harvesters: fertilising the rangelands', *Rangeland Journal*.

Chapter 24

Corbett, L (1995), *The dingo in Australia and Asia*, UNSW Press, Sydney.
Corbett, LK & Newsome, AE (1987), 'The feeding ecology of the dingo III. Dietary relationships with wildly fluctuating prey populations in central Australia: An hypothesis of alternation of predation', *Oecologia*, no. 74, pp. 215–227.
Moseby, KE, Brandle, R & Adams, M (1999), 'Distribution, habitat and conservation status of the rare dusky hopping-mouse, *Notomys fuscus*', *Wildlife Research*, no. 26, pp. 479–494.
Parkes, J, Henzell, R & Pickles, G (1996), *Managing vertebrate pests: feral goats*, Australian Government Publishing Services, Canberra.
Pople, AR, Grigg, GC, Cairns, SC, Beard, LA & Alexander, P (2000), 'Trends in the numbers of red kangaroos and emus on either side of the South Australian dingo fence: evidence for predator regulation?', *Wildlife Research*, no. 27, pp. 269–276.
Read, JL & Bowen, Z (2001), 'Population dynamics, diet and aspects of the biology of feral cats and foxes in arid South Australia', *Wildlife Research*, no. 28, pp. 195–203.
Wood-Jones, F (1923), *The mammals of South Australia*, Government Printer, Adelaide.

Chapter 25

Bowen, ZE & Read, JL (1998), 'Population and demographic patterns of rabbits (*Oryctolagus cuniculus*) at Roxby Downs in arid South Australia and the influence or rabbit haemorrhagic disease', *Wildlife Research*, no. 25, pp. 655–662.

Ceballos, G, Pacheco, J & List, R (1999), 'Influence of prairie dogs (*Cynomys ludovicianus*) on habitat heterogeneity and mammalian diversity in Mexico', *Journal of Arid Environments*, no. 41, pp. 161–172.

Coman, B (1999), *Tooth & Nail: The story of the rabbit in Australia*, Text Publishers, Melbourne.

Cooke, BD (1977), 'The rabbit in inland Australia', *Australian Mammal Society Bulletin*, no. 3, pp. 17–18.

Hollands, D (1984), *Eagles, Hawks and Falcons of Australia*, Nelson, Melbourne.

Jones, CG, Lawton, JH & Shachak, M (1994), 'Organisms as ecosystem engineers', *Oikos*, no. 69, pp. 373–386.

Lange, RT & Graham, CR (1983), 'Rabbits and the failure of regeneration in Australian arid zone Acacia', *Australian Journal of Ecology*, no. 8, pp. 377–381.

Parer, I (1977), 'The population ecology of the wild rabbit (*Oryctolagus cuniculus*) in a Mediterranean-type climate in New South Wales', *Australian Wildlife Research*, no. 4, pp. 171–205.

Robinson, AC (1992), 'Perenties, predators and prey', *South Australian Naturalist*, no. 67, pp. 30–34.

Chapter 26

Eldridge MDB, Gee, P & Gee, I (1994), 'Identification of a rock wallaby population from the Davenport Ranges, Central South Australia, as *Petrogale lateralis* Macdonnell Ranges race', *Australian Mammalogy*, no. 17, pp. 125–128.

Gee, P, Gee, I & Read, JL (1996), 'An annotated bird list from the Davenport Ranges, South Australia', *South Australian Ornithology*, no. 32, pp. 76–81.

Kinnear, JE, Onus, ML and Sumner, NR (1998), 'Fox control and rock-wallaby population dynamics – II. An update', *Wildlife Research*, no. 25, pp. 81–88.

Moseby, K, Read, JL, Gee, P & Gee, I (1998), 'A study of the Davenport Range Black-footed Rock Wallaby colony and possible threatening processes', *Final Report to Wildlife Conservation Fund*, SA Dept of Environment, Kensington.

Chapter 28

Badman, FJ (1999), 'The Lake Eyre South Study: Vegetation', *Lake Eyre South Monograph Series*, vol. 2, Royal Geographical Society of South Australia, Adelaide.

Coman, BJ & McCutchan, J (1994), *Predator exclusion fencing for wildlife management in Australia*, Australian Nature Conservation Agency, Canberra.

Owens, HM & Read, JL (1999), 'Mammals of the Lake Eyre South region', *Lake Eyre South Monograph Series*, Vol. 1, Part 2, Slater WJH (ed.), RGSSA, Adelaide.

Read, JL (2003), 'Are miners the bunnies or the bilbies of the rangelands?', *Rangeland Journal*, no. 25, pp. 172–182

Acknowledgements

Alan Andersen, Frank Badman, Peter Bailey, Rita Borda, Mike Bull, Peter Catling, Pete Copley, Steve Delean, Steve Green, Keith Greenfield, Lesley Hughes, Mark Hutchinson, Craig James, Kelli-Jo Kovac, Clive Minton, Darren Niejalke, Helen Owens, Rachel Paltridge, Andrew Phillips, Winston Ponder, Nick Reid, Mike Tyler, Max Waterman, Mike Worby and all of the characters named in this book (and many others not identified individually) provided essential technical advice, or contributed to my understanding and appreciation of the outback and the issues that it faces. Special thanks are singled out for Peter Bird, Brian Cooke, Phil Gee, Gordon Grigg, Chris Harvey, Katherine Moseby, David Paton and Hugh Possingham for advice on drafts of particular chapters. John Fewster produced the map, Steve Grimwade did a great job with the editing and Richard Yeeles, Jenny Read, Lindsay Moseby, Heather Fletcher and Vicki Kalgovas proofread various chapters and helped to prune and focus my rambling thoughts.

Even though most of the work documented in this book and its production were conducted in my own time, I am indebted to WMC and in particular to my managers Richard Yeeles, Mark Edebone, Steve Green, Jim Hondros and Barry Middleton for encouraging or allowing me to pursue my interests, and for being receptive to innovations and ideas that assisted in their stated aim of sensitively managing the environment.

I would never have spent as long at Roxby or in the outback had it not been for the companionship and good times spent with many mates who ensured that life was never tedious and that no weekends were wasted. Finally my extreme gratitude and love goes to Katherine, who was the driving force and most detailed critic behind several of the stories in this book, and who patiently supported the time-consuming process of putting it all together.